EAST ANGLIAN ARCHAEOLOGY

Middle Saxon
Animal Husbandry
in East Anglia

by Pam Crabtree

East Anglian Archaeology
Report No.143, 2012

Archaeological Service
Suffolk County Council

EAST ANGLIAN ARCHAEOLOGY
REPORT NO. 143

Published by
Suffolk County Council Archaeological Service
Economy, Skills and Environment
9–10 The Churchyard
Shire Hall
Bury St Edmunds
Suffolk IP33 2AR

in conjunction with
ALGAO East
http://www.algao.org.uk/cttees/Regions

Editor: Sue Anderson
EAA Managing Editor: Jenny Glazebrook

Set in Times Roman by Jenny Glazebrook using Corel Ventura™
Printed by Henry Ling Limited, The Dorset Press

ISBN 978 0 9568747 1 9

East Anglian Archaeology was established in 1975 by the Scole Committee for Archaeology in East Anglia. The scope of the series expanded to include all six eastern counties and responsibility for publication passed in 2002 to the Association of Local Government Archaeological Officers, East of England (ALGAO East).

For details of *East Anglian Archaeology*, see last page

This volume was published with the aid of funding from English Heritage

Cover illustration:
Male red deer skull from Brandon. The antlers have been removed by chopping and sawing (*photo by Dr Krish Seetah*)

Contents

List of Plates

List of Figures

List of Tables

Contributor

Pam J. Crabtree
Department of Anthropology, Center for the Study of
Human Origins, New York University

Acknowledgements

This project would not have been possible without the help of many people and agencies. I want to thank the US National Science Foundation (Grant BNS77-08141), the Wenner-Gren Foundation for Anthropological Research (Grant 3267) and the Fulbright-Hayes Program for the support of my initial research at West Stow. I would also like to thank Mark Maltby and the late Jennie Coy who were wonderful sources of support during my time at the University of Southampton. I am profoundly grateful to Andrew Tester and Keith Wade for allowing me to study the faunal collections from Ipswich, Wicken Bonhunt and Brandon, and I am grateful to the US National Endowment for the Humanities and to English Heritage for their support of these projects. Rosie Luff allowed me to use the comparative collections housed in the Cambridge University Faunal Remains Unit, and I am grateful for her assistance throughout this project. I would also like to thank Jude Plouviez and Catherine Hills for allowing me to examine the fauna from Icklingham.

The preparation of this manuscript would not have been possible without the help of many people. Michael Campana provided technical assistance on the Principal Components Analysis and read an earlier draft of this manuscript. Kelila Jaffe read an early draft of the text while we were excavating at the Moku'ula site in Maui. Doug Campana has been involved with the Brandon project from the start, and he patiently produced all the graphics for this volume. Catherine Hills suggested several useful papers that had not crossed the pond when I wrote the first draft of this manuscript. Andrew Tester and Keith Wade of Suffolk County Archaeological Service provided the plans of Brandon, Ipswich and Wicken Bonhunt that appear in this volume.

Many colleagues have also provided me with comparative material. I would like to thank P. Baker, M. Maltby, I. Baxter, S. Archer, C. Ingram, L. Higbee, J.-H. Yvenic and V. Chaulet for sharing their data with me. I especially want to thank Pat Stevens for her hard work on the identification of the Ipswich and Wicken Bonhunt mammal remains and Don Bramwell for the identification of the bird remains from these sites. Finally, I would like to thank the two anonymous peer reviewers who provided many useful suggestions for improvement. The errors, of course, are all mine.

Preface

When I started work on my PhD research on the Early Saxon faunal remains from the site of West Stow in Suffolk (S.E. West 1985), I was particularly interested in the transition from Roman Britain to Anglo-Saxon England. Long-term programs of excavation at sites such as Winchester had shown that many of the Roman towns had lost much of their urban character by the 5th century AD. While Wroxeter remains an important exception, with construction and occupation continuing into the 6th and 7th centuries (White and Barker 1998), most of the other Roman towns and cities in eastern England appear to have suffered serious economic decline and loss of population by the mid-5th century. As a postgraduate student, I was interested in the question of whether one might be able to see some degree of continuity between Roman Britain and Anglo-Saxon England in patterns of animal husbandry, particularly in the countryside. I was interested in whether the Anglo-Saxons preserved some of the animal size improvement that has been attributed to the Romans (see Albarella *et al.* 2008) and whether the Romans and Anglo-Saxons raised, butchered and consumed their animals in similar ways.

With 20/20 hindsight, I do think that the Anglo-Saxons maintained some of the Roman size improvement (see Chapter 5), but that Roman animal husbandry and Anglo-Saxon animal husbandry were focused on very different goals (Crabtree 1991). More importantly, I realised that the critical period of transition between the ancient and the medieval world was not the 5th century. Many of the fundamental changes that laid the foundations for medieval civilisation in Britain actually took place during the 'long eighth century' (Hansen and Wickham 2000). These changes include the rebirth of towns, the widespread adoption of Christianity, the rise of the Anglo-Saxon kingdoms, the beginnings of the open field systems (Oosthuizen 2007), the expansion of regional and long distance trade (Loveluck and Tys 2006), and the establishment of rural estate centres (see, for example, Lucy *et al.* 2009). The critical question for archaeozoologists is: how did all these social, political, and economic changes affect patterns of Anglo-Saxon animal husbandry and hunting?

The answer to this question is not only important for our understanding of Anglo-Saxon England, it is also crucial to our understanding of the origins and development of complex societies in many regions of the world. In the mid-

20th century, V. Gordon Childe (1936; 1950) argued that the social surplus, a surplus of agricultural and pastoral products that could be mobilised by emerging political leaders, was critical to the development of complex, urban societies around the globe. The questions that Anglo-Saxon archaeozoologists need to address are: i) is there archaeological evidence for the production of surplus animal products? and ii) if so, how was this surplus mobilised by the elite members of Anglo-Saxon society for their own advantage? These are fundamental questions of political economy.

With these 'big picture' questions in mind, I was honoured to have the opportunity to study the faunal remains from the Middle Saxon site of Brandon in western Suffolk. Like West Stow, Brandon is located in the Breckland region of East Anglia, an area that has traditionally been associated with sheep husbandry. Comparisons between the West Stow and Brandon faunal data would allow me to understand how the political and economic changes of the 7th and 8th century affected animal husbandry practices in the Brecklands of north-western Suffolk. A report on the fauna from Brandon has been included as part of the forthcoming Brandon site report (Crabtree and Campana n.d.).

After the initial report on the Brandon faunal assemblage was completed in the early 1990s. I was given the opportunity to work with Patricia Stevens on the animal bone remains from the Middle Saxon settlement of Wicken Bonhunt in Essex and the multi-period Saxon and Early Medieval town of Ipswich in eastern Suffolk. Pat Stevens completed the initial identifications, along with Don Bramwell who identified the bird remains, and I worked to turn these data into archive and publication reports for English Heritage.

The faunal material from Wicken Bonhunt was interesting to me because it was derived from a wealthy Middle Saxon estate centre, but one that was located in a very different environment from Brandon. Wicken Bonhunt was one of the earliest Middle Saxon settlements to have been excavated in eastern England. Historical evidence from the 11th century indicates that the area had substantial quantities of woodland that would have supported large herds of swine (see Chapters 3 and 4).

The faunal remains from Ipswich were also of great interest, because Ipswich was one of the 'wics' or emporia,

proto-urban settlements that were centres of craft production and of regional and international trade. Hodges (1982) had suggested that these *emporia*, both in Britain and on the continent, had played a crucial role in state formation in early medieval Europe. They would have been home to traders and craftsworkers who were engaged in activities other than food procurement. The question of how these *emporia* were supplied with food is critical to understanding how they functioned. Important foundational work on faunal assemblages from other British *emporia* had already been carried out by Jennifer Bourdillon at Hamwic (Bourdillon and Coy 1980, Bourdillon 1988), by Terry O'Connor (1991) at York, and by Barbara West (1989) and Kevin Rielly (2003) at London. The fundamental questions were: i) how were the craft specialists and traders at these sites provisioned with food? and ii) did the need for urban provisioning lead to economic changes in the Middle Saxon countryside? In addition, the faunal collection from Ipswich included substantial assemblages from the Late Saxon and Early Medieval periods, which allowed us to ask questions about long-term changes in urban provisioning in East Anglia (see Crabtree 2012a for a more detailed discussion of changes through time in the fauna from Ipswich).

When I first analysed the animal bone remains from Brandon, Ipswich and Wicken Bonhunt, I had hoped that the results would be published relatively rapidly as part of the site reports for these three projects. All three fell victim to the inevitable post-excavation delays. The Brandon report will be published in the near future, but the publication of Wicken Bonhunt and Ipswich is still delayed. In 2006, I wrote a proposal to *East Anglian Archaeology* for a volume that would bring together the results of the analyses of the faunal assemblages from these three sites. The goal of this volume is not to replace the individual site reports. Rather it is designed to show how the data from these three very different Middle Saxon settlements can shed light on broader patterns of animal use and urban provisioning in the Middle Saxon period. The volume is also designed to present some of the basic data from all three sites so that they can be part of the dialogue about the nature of Middle Saxon England.

Summary

This report describes the results of a comparative study of three large Middle Saxon faunal assemblages from eastern England. They include the animal bone remains from the Middle Saxon estate centres of Brandon in western Suffolk and Wicken Bonhunt in north-western Essex and the faunal remains that were recovered from a number of Middle Saxon sites within the town of Ipswich. At that time Ipswich served as an *emporium* or 'wic', a centre of craft production and regional and international trade. All three sites produced large faunal assemblages that were analysed using standard archaeozoological methods. Individual bones were identified to species and body part; the bones were examined for traces of butchery and pathology; ages at death were determined on the basis of dental eruption and

wear and epiphyseal fusion of the long bones; and measurements were recorded when possible.

Species ratios, mortality profiles and osteometric data suggest that the inhabitants of Brandon were engaged in specialised wool production. Unlike most other Anglo-Saxon sites, the Middle Saxon features at Wicken Bonhunt produced large numbers of pig bones. The residents of the site may have been engaged in large-scale pork production, and the limited evidence from the late 6th-to-7th century features at the site suggest that specialised pork production may have begun at the site in the later part of the Early Saxon period. Brandon and Wicken Bonhunt also produced rich assemblages of wild birds, including water birds and waders. The Middle Saxon sites from Ipswich yielded a much less diverse bird assemblage. The inhabitants of Ipswich appear to have been provisioned with beef and mutton from the surrounding countryside, but the ageing data indicate that some pigs may have been raised within the town itself. The results are compared to the faunal assemblages that have been recovered from other Early and Middle Saxon sites in eastern England.

Résumé

Ce rapport décrit les résultats d'une étude comparative portant sur trois grands ensembles faunistiques de la période saxonne moyenne à l'est de l'Angleterre. Ces ensembles comprennent d'une part les restes d'ossements animaux provenant des domaines de Brandon à l'ouest du Suffolk et de Wicken Bonhunt au nord-ouest de l'Essex, et d'autre part, les restes faunistiques retrouvés dans plusieurs sites de la ville d'Ipswich. Ces sites, ainsi que les domaines précédemment nommés, datent de la période saxonne moyenne. À cette époque, Ipswich était un centre de production artisanale et de commerce tant régional qu'international; il remplissait ainsi un rôle d'*emporium* ou de «*wic*». Ces trois sites contenaient de grands ensembles faunistiques qui ont été analysés selon des méthodes archéozoologiques standard. Les os distincts ont été identifiés en fonction de l'espèce et de la partie du corps concernées; les recherches ont également porté sur les traces de découpe et sur les indices de pathologie. Les âges de la mort ont été déterminés en analysant l'éruption et l'usure des dents ainsi que la fusion épiphysaire des os de grande taille. Des mesures ont également été prises dans la mesure du possible.

La proportion des espèces, les profils de mortalité et les données ostéométriques suggèrent que les habitants de Brandon s'étaient spécialisés dans la production de laine. On trouve un grand nombre d'os de porc dans le site de la période saxonne moyenne de Wicken Bonhunt, à la différence de la plupart des autres sites anglo-saxons. Il est possible que les résidents du site se soient lancés dans la production de porc à grande échelle. En outre, le faible nombre de preuves datant de la fin du 6ème siècle et du 7ème siècle donne à penser que la production exclusive de porc a peut-être commencé sur le site à la fin de la première période saxonne. On a également découvert à Brandon et à Wicken Bonhunt de riches ensembles d'oiseaux sauvages comprenant des oiseaux aquatiques et des échassiers. Les ensembles d'oiseaux trouvés dans les sites d'Ipswich à la période saxonne moyenne sont nettement moins diversifiés. Les habitants d'Ipswich étaient apparemment approvisionnés en viande de bœuf et de mouton venant de la campagne environnante mais les données relatives aux âges indiquent que certains porcs ont peut-être été élevés à l'intérieur de la ville. Les résultats obtenus sont comparés aux ensembles faunistiques qui ont été retrouvés sur d'autres sites de l'est de l'Angleterre datant des périodes saxonnes initiale et moyenne.

(Traduction: Didier Don)

Zusammenfassung

Der Bericht beschreibt die Ergebnisse einer vergleichenden Studie zu drei großen Faunenkomplexen aus der Mitte der angelsächsischen Zeit in Ostengland. Gefunden wurden Reste von Tierknochen in den damaligen Verwaltungszentren Brandon im Westen von Suffolk und Wicken Bonhunt im Nordwesten von Essex sowie Tierreste, die von mehreren mittelangelsächsischen Fundstellen in der Stadt Ipswich stammen. Ipswich war damals ein *Emporium* (oder «*wic*») — ein Zentrum der Handwerksproduktion und des regionalen und internationalen Handels. Alle drei Fundstellen wiesen umfangreiche Faunenreste auf, die mit Hilfe standardisierter archäozoologischer Verfahren analysiert wurden. Einzelne Knochen konnten bestimmten Tiergattungen und Körperteilen zugeordnet werden. Die Knochen wurden auf Schlachtspuren und pathologische Befunde untersucht, das Alter zum Todeszeitpunkt wurde anhand von Zahndurchbruch und Zahnabrieb sowie aufgrund der Epiphysenfugen der Röhrenknochen bestimmt, und wo immer möglich wurden Messungen aufgezeichnet.

Das zahlenmäßige Verhältnis lässt zusammen mit den Mortalitätsprofilen und den osteometrischen Daten der verschiedenen Tiergattungen darauf schließen, dass die Einwohner von Brandon speziell der Wollerzeugung nachgingen. Die mittelangelsächsischen Fundkomplexe von Wicken Bonhunt wiesen anders als die meisten anderen angelsächsischen Fundstellen zahlreiche Schweineknochen auf. Möglicherweise erzeugten die Bewohner in großem Umfang Schweinefleisch, wobei das begrenzte Fundmaterial aus dem späten 6. und dem 7. Jahrhundert an diesem Ort nahelegt, dass die Schweinefleischerzeugung bereits gegen Ende der frühangelsächsischen Zeit eingesetzt haben könnte. In Brandon und in Wicken Bonhunt wurden ferner umfangreiche Wildvogelreste, darunter von Wasser- und Watvögeln, gefunden. Die Vogelreste der mittelangelsächsischen Fundstellen in Ipswich waren weit weniger vielfältig. Wie es scheint, wurden die Menschen in Ipswich mit Rind- und Hammelfleisch aus der Umgegend versorgt, allerdings deutet die Altersdatierung darauf hin, dass möglicherweise auch in der Stadt selbst einige Schweine gehalten wurden. Die Ergebnisse werden mit Faunenresten von anderen früh- und mittelangelsächsischen Stätten in Ostengland verglichen.

(Übersetzung: Gerlinde Krug)

Chapter 1. Introduction

I. Introduction

The Middle Saxon period (conventionally dated *c.* AD 650–850) was an era of significant social, political, and economic change in Anglo-Saxon England. Important transformations that took place during this period include the adoption of Christianity and the spread of monasticism (Blair 2005), the political consolidation of the Anglo-Saxon kingdoms, and the foundation of the *emporia*, the earliest towns in post-Roman northern Europe (see, for example, Hodges and Hobley 1988). Many of the Early Saxon settlements, such as West Stow in western Suffolk, were abandoned, and new settlements, including the *emporium* of Ipswich and the rural estate centre of Brandon, were established. Although written documentation is sparse, archaeological data indicate that significant changes in trade and craft production also took place during the 'long eighth century' (Hansen and Wickham 2000). Since animal husbandry played a critical role in Anglo-Saxon economic and ritual life, it is reasonable to ask whether these political and economic changes were accompanied by alterations in the ways that animals were raised and in the ways their primary and secondary products were distributed and consumed.

In terms of settlement patterns and landscape usage, Reynolds (2005, 115) has argued that, 'It has become increasingly clear in archaeological terms that the seventh century was a pivotal era in England'. During the Early Saxon period, villages such as West Stow represented 'unbounded settlements, set within existing field systems' (Reynolds 2005, 117). At West Stow, many of the final features are boundary ditches, suggesting a fundamental reorganisation of space, and possibly property ownership, within the village. The 7th century saw the rise of a true hierarchy of settlement in Anglo-Saxon England. Not only were there proto-urban settlements like Ipswich, but wealthy rural estate centres also appeared, including Brandon in Suffolk (Carr *et al.* 1988), Wicken Bonhunt in Essex (Wade 1980), Flixborough in Lincolnshire (Loveluck and Atkinson 2007, Dobney *et al.* 2007) and Ramsbury in Wiltshire (Haslam 1981). Some of these settlement changes were undoubtedly related to the emerging Anglo-Saxon royal houses' need to regulate commerce and collect taxes. In Middle Saxon England, administration took place through a series of royal vills. The estates paid a render of tax, known as the *feorm*. The king and his court were itinerant, essentially eating their way through the kingdom. Each district was required to contribute to the support of the king and his retinue (Hooke 1998, 50). In a system of staple finance (Brumfiel and Earle 1987) such as this one, animal products may have served as an important source of revenue. The site of Higham Ferrers in Northamptonshire (Hardy *et al.* 2007) appears to be such a food-rent collection centre.

This volume will examine the role that animal husbandry played in the economy of Middle Saxon East Anglia. The study is based on the comparative analysis of the faunal remains from three archaeological projects, Ipswich and Brandon in Suffolk and Wicken Bonhunt in Essex (Figure 1.1). Ipswich is one of the Middle Saxon *emporia* or *wics*. It served as a centre of craft production and regional and international exchange. Faunal remains from this site can reveal how the emerging urban centre was provisioned with meat and other animal products. Brandon and Wicken Bonhunt are both sites that probably served as rural estate centres, and these sites can shed light on the nature of Middle Saxon animal production and the ways that it may have changed on response to opportunities for trade and demands for taxes and food-renders. Extensive excavations at these three sites have produced substantial faunal assemblages that can be used to reconstruct Middle Saxon animal husbandry.

The theoretical approach adopted in this volume is predominantly palaeoeconomic (*cf.* Higgs 1975); the goal is to use faunal remains to reconstruct Middle Saxon animal husbandry practices, hunting patterns and diet. However, the approach taken here differs from the more traditional palaeoeconomic approaches of the 1970s. Today most archaeologists recognise that economic practices cannot be divorced from social, political, and religious life. My goal is to use these economic data to shed some light on the broader social, political, and economic changes that took place between about 650 and 850 AD.

This chapter will provide a brief introduction to the archaeology of the Brandon, Wicken Bonhunt and Ipswich sites, followed by an overview of the methods used to identify and analyse the faunal remains from these three archaeological projects. The following chapter will provide a review of the archaeozoology of the Iron Age, Roman and Early Saxon periods in East Anglia, based primarily, but not exclusively, on the fauna from West Stow (S.E. West 1985, 1990) and Icklingham (West and Plouviez 1976).

II. Brandon

The Staunch Meadow Brandon site (BRD 018) is located in north-western Suffolk. The Suffolk County Archaeological Unit, under the direction of Andrew Tester and Bob Carr (Carr *et al.* 1988) carried out eight seasons of excavation at the site between 1979 and 1988. Approximately one-third of the site, an area of 13,000m², was excavated in advance of the construction of playing fields. The excavation revealed 34 post-built timber buildings, plus fence lines, pits, ditches, hearths, and a church and cemetery. There was a waterfront industrial area on the north side of the site (Figure 1.2). Since the site was never ploughed, the Middle Saxon remains, including the fauna, were exceptionally well preserved.

Artefactual remains, such as silver pins, indicate that Brandon was a wealthy community, and the presence of a stylus and ink wells suggests that at least some members of the community were literate. Brandon was clearly a Christian community, and it may have served as a monastic double house for part, if not all, of its existence (Andrew Tester, pers. comm.). The site appears to have

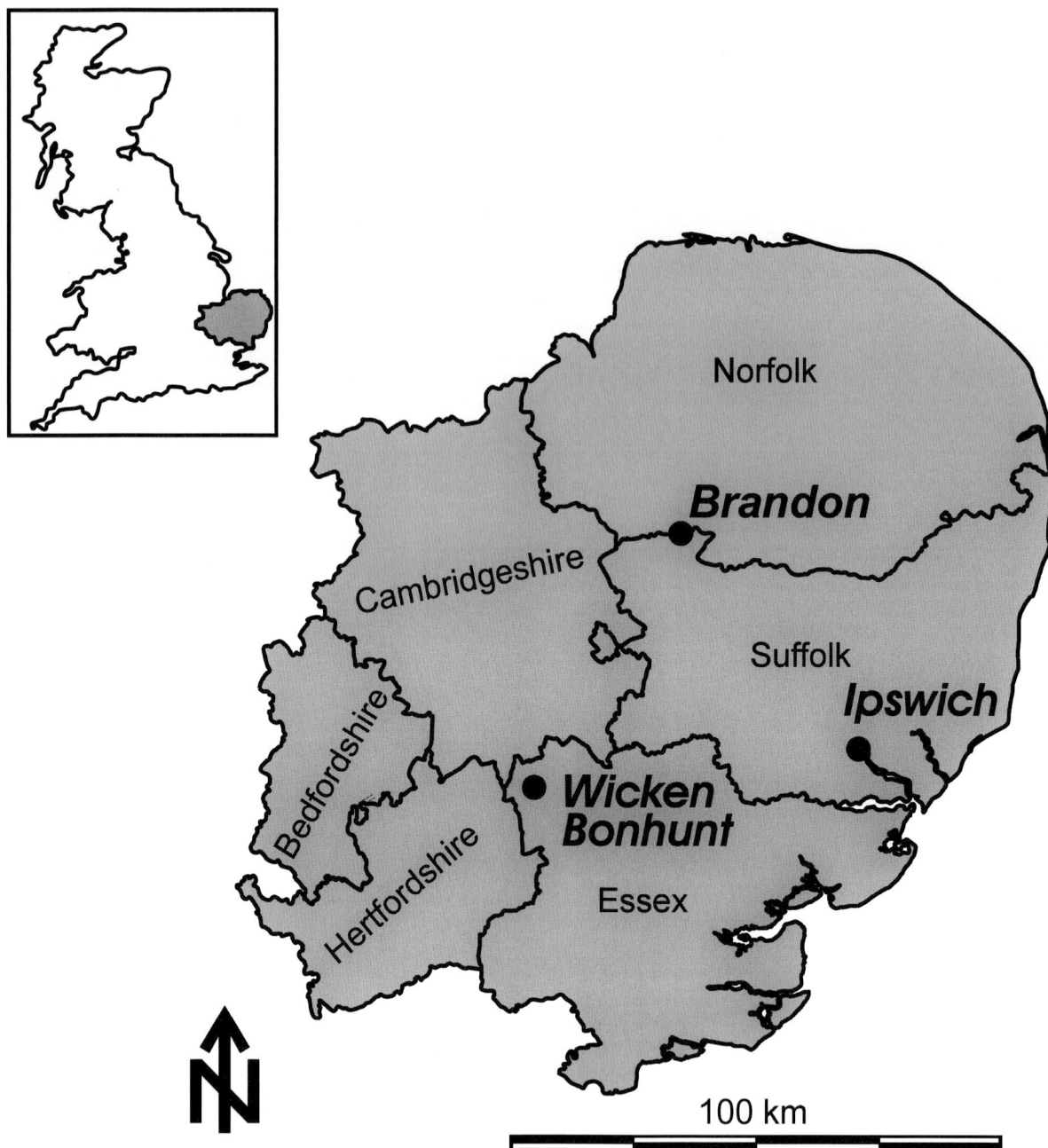

Figure 1.1 Map of East Anglia showing the locations of the Brandon, Wicken Bonhunt and Ipswich sites

been initially occupied around AD 650. Recent analyses of the ceramics and other datable artefacts indicate that the site was abandoned around 850, possibly as a result of the Danish invasions of East Anglia (Andrew Tester, pers. comm.).

The excavations at Staunch Meadow Brandon produced a huge collection of faunal material, including over 158,000 mammal and bird bones. Most of the fauna came from the general culture layer, about 10–15cm thick, that covered the site. Very few animal bones could be assigned to one of the chronological subphases that have been defined within the Middle Saxon period.

III. Wicken Bonhunt

The parish of Wicken Bonhunt lies 64km north-north-east of London and 32km south-south-east of Cambridge in the north-west corner of Essex. A multi-period settlement was uncovered at Bonhunt Farm which lies to the east of the parish on a south-facing slope of the River Cam (Figure 1.3). The site was discovered in 1967 after a field adjacent to a small Norman chapel and redundant farm buildings was ploughed. Considerable quantities of pottery were recovered, ranging in date from the Roman period to the 13th century. The ceramic assemblage included Middle Saxon Ipswich ware and grass-tempered pottery. The following year, bulldozing revealed

Figure 1.2 Plan of Brandon

additional Anglo-Saxon material associated with possible habitation sites.

Rescue excavations took place at the site between 1971 and 1973 under the direction of Keith Wade and Andrew Rogerson (Wade 1980). Excavation was carried out by stripping the topsoil down to the natural subsoil and then excavating the exposed features by hand. Although the aim of the excavation was to open up as large an area as possible, excavation was limited to those areas of the site that were directly threatened by farming operations. All the archaeological material was hand-collected, and no systematic sieving programme was carried out at the site.

Flotation samples were taken from several Roman, Middle Saxon and 11th-century features.

The main area of excavation revealed ditches and pits belonging to the Roman period, but no structures were present. Although ceramic evidence indicates Early Saxon occupation, no structures dating to this period were identified.

The Middle Saxon features include boundary ditches, at least 28 structures, and two wells. A large channel to the south of the site may be the leat of a watermill. Although there is no direct evidence for the function of any of the Middle Saxon structures, some of the buildings are likely

MIDDLE SAXON

O

C

P Q
W
N
V
F
T
X Y
G
J
D E
AD
AC H
U O M
R AA Z
L
AB
AH
B

S

? LEET

0   20m

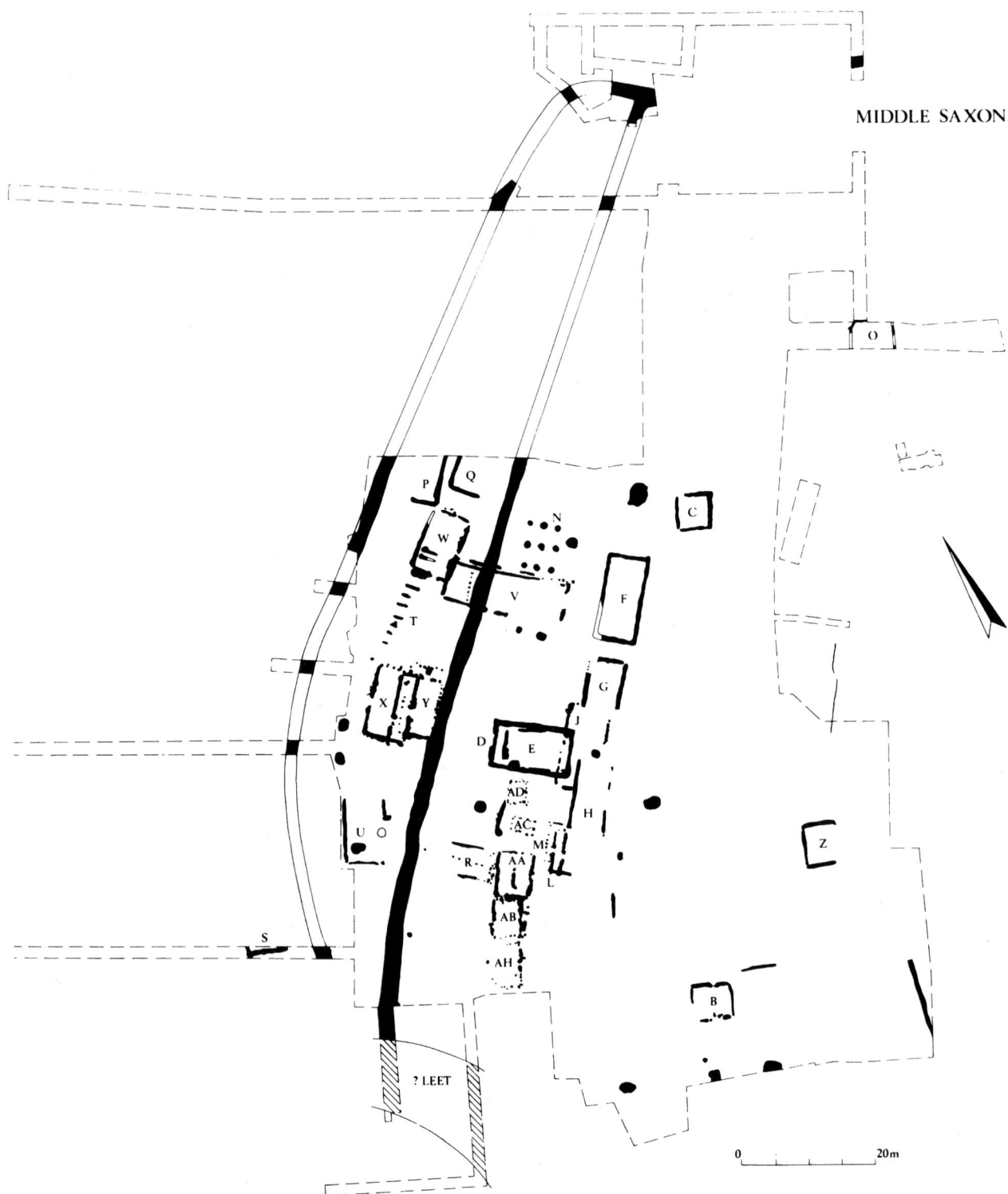

Figure 1.3 Plan of Wicken Bonhunt

to have served as workshops, barns and byres (Wade 1980, 96). There appear to be at least two major phases of building activity at the Bonhunt site (Wade 1980, 96). The location of the settlement boundaries suggests that between approximately one-third and one-half of the site has been excavated. The building alignments indicate that their layout was deliberately planned, suggesting a high degree of organisation (Wade 1980, 98).

More than 100,000 animal bones and fragments were recovered from the Middle Saxon features at Wicken Bonhunt. Approximately 70,000 of those were recovered from a large boundary ditch. The remainder of the Middle Saxon fauna came from ditches, wells and pits.

4

IV. Ipswich

Systematic rescue excavations begun in 1974 have shown that Ipswich was founded in the 7th century, probably by the East Anglian royal house. The royal burial ground lies approximately 15km north-east at the site of Sutton Hoo (Carver 2005). During the Middle Saxon period Ipswich served as a site of both industrial production and international trade. Imported pottery indicates that trade focused on the Rhineland, Flanders and northern France (Scull 2009, 318). Craft activities included the production of Ipswich ware ceramics which were turned on a slow wheel and kiln-fired, as well as weaving, metallurgy, and bone- and leather-working.

Recent excavations in Ipswich have identified an early 7th-century settlement located on the north bank of the River Orwell that predates the production of Ipswich ware (Scull 2009). From the beginning, this settlement produced evidence for trade and craft production. A high proportion of the pottery was imported wares, suggesting that this was a special-purpose settlement designed to channel and control exchange (Scull 2009, 314). It was probably established by the East Anglian royal house, although Kentish or continental authorities may also have played a role. However, the settlement was set within an agricultural landscape and was neither urban nor proto-urban at this time.

Ipswich expanded rapidly in the 8th century, possibly beginning as early as AD 680–700, to cover an area of about 50ha. The town expanded northward onto previously unoccupied heathland, and a gridiron street plan was laid out. Ipswich ware may initially have been produced some time around AD 700–720 (Blinkhorn 1999), although this is still a subject to debate. The site was initially fortified in the 10th century, probably under the Danish occupation. The town expanded little during the Late Saxon period, when it continued to serve as a centre of craft production and exchange. By the Late Saxon period, however, trade was more regional than international. Ipswich continued as a craft production centre and market town for east Suffolk into the medieval period (Wade 1988; 1993; 2000).

Animal bones were recovered from sixteen sites that were excavated between 1974 and 1988. The sites range in date from early Middle Saxon through Early Medieval. Although sizeable faunal assemblages were recovered from early Late Saxon (late 9th century), middle Late Saxon (10th century), and Early Medieval contexts in Ipswich, this report will focus on the animal bones recovered from the eleven sites that produced Middle Saxon faunal material (Table 1.1). A list of the faunal remains recovered from each of these sites is included in the Appendix. The locations of these sites are shown on Figure 1.4. All the faunal remains recovered from Sites 11–19 and Site 25 were recorded and analysed. Due to the mass of faunal remains recovered from Sites 27–29, these bones were sampled by the excavator prior to analysis.

V. Materials and Methods

The bird and mammal remains from Brandon were identified by the author and Douglas Campana (Crabtree and Campana n.d.) during 1990–91 using the comparative collections housed in the former Cambridge University Faunal Remains Unit, headed by Dr R. Luff. A small number of unusual specimens were identified using the collections housed at the Fitzwilliam Museum of Zoology at Cambridge University and the American Museum of Natural History in New York.

Basic identification included species, anatomical element, side, portion, and degree of fragmentation. Degree of fragmentation was recorded using the old English Heritage system. Bones were scored as less than half, roughly half, more than half, and complete or nearly complete. Higher order taxa, such as sheep/goat, large ungulate ('cattle sized'), and small artiodactyl ('sheep-sized') were used to classify bones that could not be identified to species. The use of these higher order taxa follows Crabtree (1990a, 5). Following the recommendations of Lyman (2008, 81), species ratios were calculated using NISP (number of identified specimens per taxon).

The Brandon bones were measured following the recommendations of von den Driesch (1976). Only mature adult bones were measured. Most measurements were taken to the nearest 0.1 mm using a Helios dial caliper. Greatest lengths of the large mammal long bones were taken using an osteometric board and were measured to the nearest 0.5mm. Withers' heights were calculated from these measurements following von den Driesch and Boessneck (1974).

Estimates of ages at death for the domestic mammals were based on both epiphyseal fusion of the long bones (Silver 1969) and dental eruption and wear (Payne 1973; Grant 1982). Mandible Wear Stages were calculated for cattle, sheep, and pig mandibles, following Grant (1982), and sheep and goat mandibles were also grouped into age classes following Payne (1973). In addition, complete and

Site	Location and number	NISP	Comments
Site 11	Foundation Street/Star Lane (5810)	429	Evidence for antler-working
Site 13	Tower Ramparts (0802)	39	Bones recovered from pits
Site 14	Little Whip Street Site (7407)	24	Bones recovered from pit during flotation
Site 16	Bridge Street Site (6202)	1538	Antler waste, worked bones, and goat horn cores
Site 17	St Peter's Street (5202)	169	Antler waste
Site 18	Key Street Site (5901)	92	Waste from antler- and horn-working
Site 19	Shire Hall Yard Site (6904)	525	
Site 25	Foundation Street/School Lane (4801)	1017	
Site 27	Foundation Street/Wingfield (4601)	2184	Only a sample of the recovered fauna was analysed
Site 28	St Peter's Street (5203)	3939	Only a sample of the recovered fauna was analysed
Site 29	Buttermarket (Greyfriars Rad.) (3104)	574	Only a sample of the recovered fauna was analysed

Table 1.1 Middle Saxon sites from Ipswich

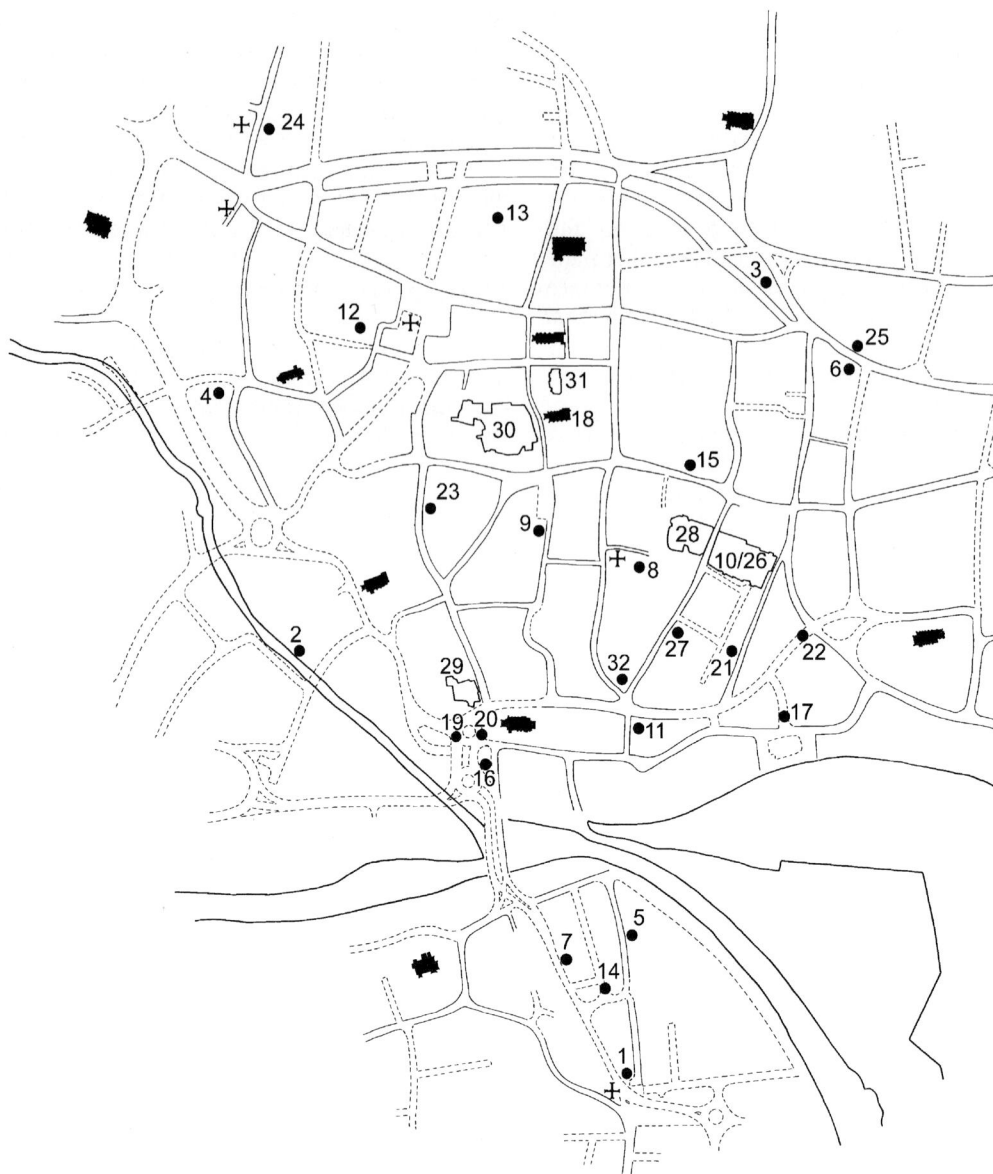

1. Vernon Street / Gt Whip Street, 1974 (IAS 7501)
2. Cecilia Street, 1974 (IAS 5001)
3. Old Foundry Road, 1974 (IAS 1501)
4. Elm Street, 1975 (IAS 3902)
5. Gt Whip Street, 1975 (IAS 7501)
6. St Helen's Street, 1975 (IAS 3601)
7. Vernon Street, 1975 (IAS 7402)
8. Lower Brook Street, 1975 (IAS 5502)
9. Turret Lane, 1978 (IAS 4302)
10. School Street, 1979 (IAS 4801)
11. Foundation Street / Star Lane, 1979 (IAS 5801)
12. Arcade Street, 1979 (IAS 1804)
13. Tower Ramparts, 1979/81 (IAS 0802)
14. Lt Whip Street, 1980-81 (IAS 7404)
15. Tacket Street, 1980-81 (IAS 3410)
16. Bridge Street, 1981 (IAS 6202)
17. Key Street, 1981 (IAS 5901)
18. St Stephen's Church, 1982 (IAS 3203)
19. Greyfriars Road, 1982 (IAS 5201)
20. St Peter's Street / Greyfriars Road, 1982 (IAS 5202)
21. Shire Hall Yard, 1982 (IAS 6904)
22. Fore Street, 1982 (IAS 5902)
23. St Nicholas Street, 1983 (IAS 4201)
24. St George's Street, 1983 (IAS 9802)
25. St Helen's Street, 1983 (IAS 8804)
26. School Street / Foundation Street, 1983-85 (IAS 4801)
27. Smart Street / Foundation Street, 1984 (IAS 5701)
28. Wingfield Street / Foundation Street, 1985 (IAS 4601)
29. Greyfriars Road, 1986 (IAS 5203)
30. St Stephen's Lane, 1987-88 (IAS 3104)
31. Buttermarket, 1987-88 (IAS 3201)
32. Lower Brook Street / Foundation Street, 1988 (IAS 5505)

Figure 1.4 Plan of Ipswich showing the locations of sites that produced Middle Saxon faunal remains

6

nearly complete cattle and sheep mandibles were grouped into age classes following Bourdillon and Coy (1977; 1980). This is a variation of the system first proposed by Payne (1973) for sheep and goats, but Payne's nine age classes have been reduced to six by combining the A and B (0–6 months), E and F (2–4 years), and G and H (4–8 years) stages. This was done to make the Brandon age profiles comparable to those recovered from the *emporium* of Hamwic (Anglo-Saxon Southampton) in Hampshire (Bourdillon and Coy 1977; 1980) and the Early Saxon site of West Stow in Suffolk (Crabtree 1982; 1990a).

The Brandon animal bones were also examined for traces of butchery, pathology, and bone working. A small number of the most interesting specimens were photographed, using both traditional and digital photography. All the faunal data from Brandon were entered into a specialised database called ANIMALS (Crabtree and Campana 1987).

The Ipswich and Wicken Bonhunt faunal collections have a very different history. The bird and mammal remains from Wicken Bonhunt and the mammal remains from Ipswich were initially identified by Patricia Stevens. The avifauna from Ipswich was identified by Don Bramwell. In 1992, English Heritage asked the present author to prepare Patricia Stevens' data for publication. This volume will present the results of the analysis and interpretation of the Ipswich and Wicken Bonhunt data based on Stevens' original identifications.

The original data from Ipswich and Wicken Bonhunt were collected using essentially the same methods that were used to collect the Brandon data. Bone measurements were also taken following the recommendations of von den Driesch (1976), and both dental (Grant 1982) and epiphyseal fusion (Silver 1969) data were collected for the domestic mammals. The major differences are in the distinction between sheep and goats. For Brandon, sheep and goats were distinguished following the recommendations of Boessneck *et al.* (1964). Sheep and goat mandibles and loose teeth were not distinguished, since the Halstead *et al.* (2002) paper had not been published at the time that the original research was completed. For Ipswich, however, sheep and goat post-cranial remains were rarely distinguished. As a result, the bulk of the material identified as goat came from horn cores. Since male goat horns were commonly used in horn-working industries at *emporium* sites (see, for example, Bourdillon and Coy 1980, 97, 111), the

relative importance of sheep and goat horn cores may not accurately reflect the overall proportions of sheep and goats in the faunal sample. At Wicken Bonhunt, sheep and goats recovered from a large boundary ditch were distinguished on the basis of horn cores only. In the subsequent analysis of the fauna recovered from the wells and other Middle Saxon features, sheep and goat bones were distinguished based on the criteria developed by Boessneck *et al.* (1964). This analysis was completed by Dr Simon Davis, formerly of the Ancient Monuments Laboratory.

A second difference is in the nature of the recording systems used to document the faunal data. As noted above, Brandon was recorded using the ANIMALS program, which is a flexible program designed for the PC. It allows the analyst to recombine and reanalyse different classes of data by species (or group of species), by chronological period, and by geographical location. The Wicken Bonhunt and Ipswich data were recorded using the former Ancient Monument Laboratory system (Jones n.d.). This system was not designed for the PC, and the system makes it nearly impossible to reanalyse non-metrical data. For example, it is not possible to separate the epiphyseal fusion data by site within Ipswich.

VI. Prospectus

The detailed analysis of faunal collections from Iron Age, Roman and Early Saxon sites in eastern England is presented in Chapter 2. This chapter provides the zooarchaeological background for the study of rural and urban Middle Saxon faunal collections from East Anglia. This study will begin with an overview of the animal species that were recovered at Brandon, Ipswich and Wicken Bonhunt, and their relative proportions (Chapter 3). Chapter 4 will examine the ageing data from the three sites and the implications of these data for patterns of animal husbandry and economic specialisation. The osteometric data from Brandon, Wicken Bonhunt and Ipswich will be examined in Chapter 5. These data will be used to track long-term changes in animal sizes and their implications for Middle Saxon animal economy. A final chapter, Chapter 6, will compare the East Anglian data to the faunal evidence from other Middle Saxon sites in England in order to draw some broader conclusions about animal husbandry practices and hunting patterns in Middle Saxon England.

Chapter 2. The Iron Age, Roman and Early Saxon Background to this Study

I. Introduction

The Middle Saxon landscape of East Anglia was a product of centuries and millennia of human land use and environmental change. In order to provide some long-term historical context for the analysis of the faunal remains from Brandon, Ipswich and Wicken Bonhunt, this chapter will explore the patterns of animal use that have been documented for later Iron Age, Roman and Early Saxon East Anglia. Much of the information about Iron Age and Roman periods is derived from the study of the faunal remains from the late Iron Age settlement at West Stow and the late Roman small town of Icklingham. These faunal assemblages were analysed in the same way that the Brandon assemblage was, facilitating comparisons with the Middle Saxon assemblages. However, reference will be made to other well analysed Iron Age and Roman faunal assemblages from the region, including Luff's (1993) study of the faunal remains from the Roman town of Colchester in Essex and Searjeantson's analyses of the faunal remains from the Iron Age fen-edge community at Haddenham in Cambridgeshire (Evans and Serjeantson 1988).

When the analysis of the faunal remains from the Early Saxon village of West Stow was carried out, there were very few other faunal assemblages from Early Saxon sites that had been analysed to a modern standard. Recent excavations at West Stow Visitor Centre have revealed additional structures and a substantial faunal assemblage. These animal bones were analysed in the same way as the original West Stow assemblage. However, since the fauna and the associated archaeological material are still under study they are discussed separately here. In the past 20 years, a number of other Early Saxon animal bone assemblages have been studied and published. While many of these collections are relatively small, they help to provide the economic background for the Middle Saxon faunal assemblages that are the focus of this monograph.

II. Iron Age Fauna

While West Stow is known primarily as an Early Saxon settlement, a number of Iron Age features, including a house, pits, and ditches were also recovered from the site (S.E. West 1990). The earliest Iron Age material from the Iron Age may be as early as the 3rd century BC, but most of the material dates to the early and middle part of the 1st century AD. Over 7500 animal bones and fragments were recovered from the Iron Age features at West Stow. The vast majority of the faunal remains were identified as domestic mammals, including cattle (*Bos taurus*), sheep (*Ovis aries*), goat (*Capra hircus*), pig (*Sus scrofa*), and horse (*Equus caballus*). Species ratios based on fragment counts (NISP or number of identified bones per taxon) indicate that cattle were the most common species, followed by caprines (sheep and goats), pigs, and horses (Figure 2.1).

The measurements taken on the Iron Age animal bones from West Stow are presented in detail in Crabtree (1990a). Withers height estimates for the Iron Age cattle range from 100 to 116cm, with a mean of 107cm (Crabtree 1990b, 103). These animals are comparable to the small Iron Age cattle that have been recovered from other Iron Ages sites in southern Britain. Withers height estimates for the West Stow Iron Age horses range from 110–136cm (11–13.2 hands). These ponies are similar in size to the horses that were recovered from other Iron Ages sites such as Gussage All Saints in Dorset (Harcourt 1979, 153).

Age profiles for cattle, based on dental eruption and wear, indicate that cattle of all ages were killed at Iron Age West Stow, with no concentration on either juvenile or elderly individuals. Age profiles for sheep and goats indicate that about one-third of the animals were killed during the first year of life and that over half were killed by two years of age. Most of those that survived to adulthood were killed between four and eight years of age.

The faunal data from Iron Age West Stow suggest that the site was a small, largely self-sufficient farming community whose economy was based on mixed animal husbandry. Sheep and cattle were probably used for a variety of purposes, including meat, milk, wool, and traction.

The excavation of the Haddenham V complex on the fenland edge near Ely produced a unique assemblage of wetland faunal remains dating to the Iron Age (Evans and Serjeantson 1988). In addition to cattle, sheep, and pigs, the assemblage also yielded the remains of beaver (*Castor fiber*), swan (*Cygnus olor*), and pelican (*Pelicanus crispus*), along with a range of other wild birds. Among the mammals, beaver bones actually outnumber those if pigs. These data suggest that cereal cultivation and stock-keeping based on cattle and sheep husbandry formed the basis of the domestic economy, but that beaver pelts and other fenland-specific resources such as feathers may have been collected for trade. These data suggest that the Iron Age economy may have been more complex that the West Stow data alone might suggest.

III. Roman Animal Remains

The 4th-century AD site of Icklingham in western Suffolk provides information about animal husbandry practices in East Anglia during the late Roman period. The site has been the subject of two different programmes of excavation. The first was carried out in the 1970s under Stanley West and Jude Plouviez (1976); a more recent campaign of excavation was conducted by Dr Catherine Hills of Cambridge University between 1997 and 2000. While the site of Icklingham was initially identified as a Roman villa because a bathhouse was discovered there in

Species Ratios (NISP) for Icklingham, Hills Excavation

Percent NISP

Cattle　Sheep / Goat　Pig　Horse

Figure 2.2 Species ratios (based on NISP) for the large domestic mammals from the Hills (1997–2000) excavations at Icklingham

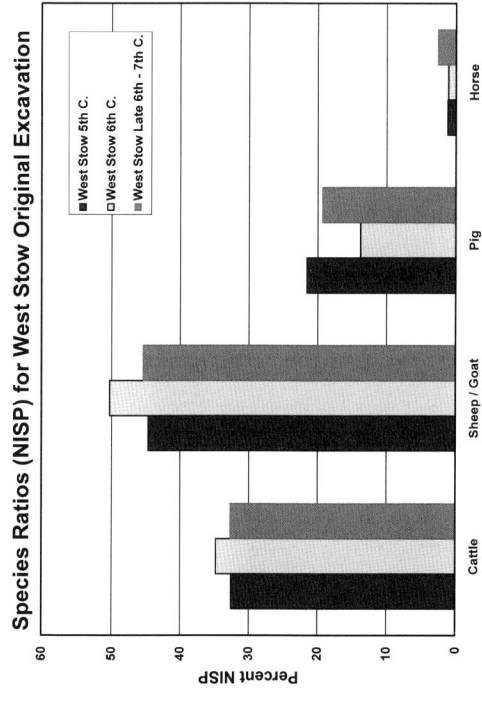

Species Ratios (NISP) for West Stow West

Percent NISP

Cattle　Sheep / Goat　Pig　Horse

Figure 2.4 Species ratios (based on NISP) for the main domestic mammals from the excavations at West Stow Visitor Centre

Species Ratios (NISP) for Iron Age West Stow

Percent NISP

Cattle　Sheep / Goat　Pig　Horse

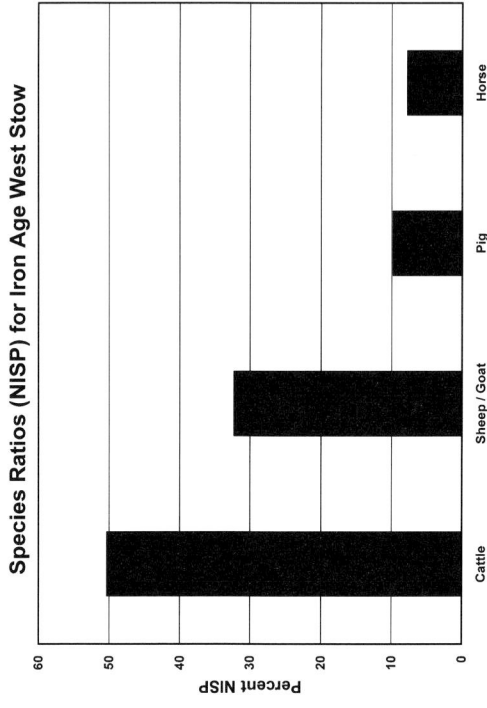

Figure 2.1 Species ratios (based in NISP) for the large domestic mammals from Iron Age West Stow

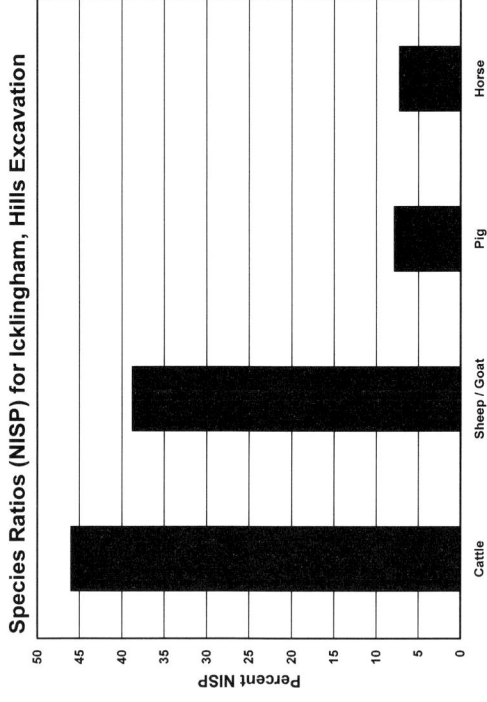

Species Ratios (NISP) for West Stow Original Excavation

Percent NISP

■ West Stow 5th C.
□ West Stow 6th C.
■ West Stow Late 6th - 7th C.

Cattle　Sheep / Goat　Pig　Horse

Figure 2.3 Species ratios (based on NISP) for the large domestic mammals from West Stow

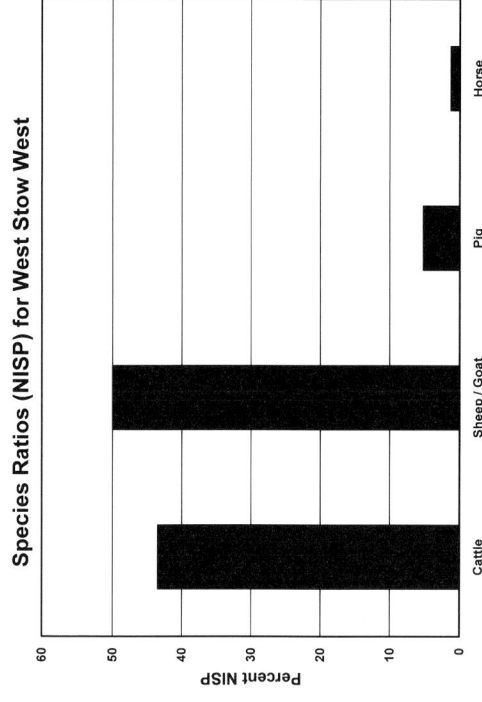

9

the mid-19th century, the recent programmes of excavation have shown that the site is actually a sprawling country market town. A study of roughly two-thirds of the fauna from the West-Plouviez excavations, about 11,800 animal bones and fragments, was carried out in 1989 (Crabtree 1991). The remainder of the material from the West-Plouviez excavations was analysed in the summer of 2008. The approximately 10,200 animal bones recovered from the 1997–2000 excavations at Icklingham have recently been analysed as well (Crabtree 2010a).

The Icklingham assemblages were dominated by the remains of cattle, followed by sheep, pigs, and horses. The species ratios for the fauna recovered from the Hills excavations (Figure 2.2), in particular, are very similar to the ratios seen from Iron Age West Stow. The kill-patterns, however, show marked differences between the Iron Age and Roman assemblages. While the Iron Age cattle were killed at all stages of life, the Roman assemblages from Icklingham include no neonatal and young juvenile cattle. In addition, the largest of the cattle from Icklingham are larger than any cattle that were recovered from Iron Age West Stow. These animals appear to represent improved Roman varieties of cattle. Larger cattle also appear in other parts of the Empire during the Roman period and in areas just outside the Empire that had trade contacts with Rome (see, for example. Tiechert 1984, MacKinnon 2010a). The cattle and other livestock from Icklingham appear to have been butchered in very standardised ways. Plate 2.1 illustrates two cattle radii showing axial chops through the proximal end made in an anterio-posterior (cranio-caudal) direction. Similar chop marks are seen on the distal humeri from Icklingham. These chop marks would have been made during the disarticulation of the forelimb, and they would also have opened up the marrow cavity. Very similar chop marks have been documented on Roman material from Winchester (Maltby 2010, 132–3). The absence of juvenile cattle suggests that the town of Icklingham was supplied with meat from the surrounding countryside, and the standardised butchery patterns indicate that the cattle were butchered by full-time specialists.

Additional evidence for animal exploitation in Roman East Anglia is provided by Luff's (1993) comprehensive study of the animal bones recovered from the 1971–85 excavations in the Roman town and fortress at Colchester, Essex. Quantitative data show relatively equal numbers of cattle sheep and pigs in most of the Roman-period deposits from Colchester (Luff 1993, 45). Most of the meat-bearing cattle bones appear to have been dumped outside the town walls, and Luff (1993, 54) suggests that this may be evidence for an organised butchery trade during Roman times. Most of the cattle consumed at Colchester were mature, with younger animals (aged 24–30 months) becoming more important in the later periods. Sheep, on the other hand, were slaughtered at young ages, and may have been kept for their milk in addition to their meat.

IV. Early Saxon Fauna

The excavation of the Early Saxon village of West Stow (S.E. West 1985) provided detailed information on animal husbandry patterns and hunting practices during the pagan Anglo-Saxon period. The West Stow village was excavated by Dr Stanley West between 1965 and 1972.

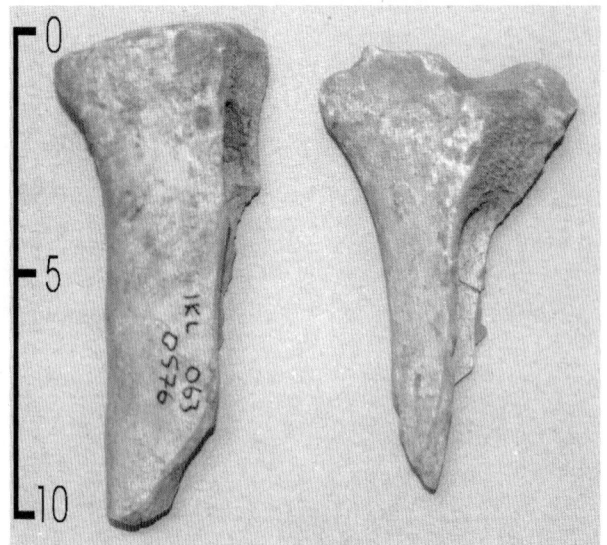

Plate 2.1 Photograph showing Roman butchery from Icklingham: two cattle radii that have been axially split

The excavations revealed sixty-nine sunken-featured buildings (SFBs or grubenhaüser) clustered around six small timbered halls. The site was occupied from approximately AD 420 to 650, and the site has been divided into three successive chronological phases, corresponding to the 5th, earlier 6th, and late 6th through 7th centuries. However, the presence of sherds of Ipswich ware in the latest features at West Stow may now indicate that the site continued into the early 8th century.

The faunal remains from West Stow have been published elsewhere in detail (Crabtree 1982; 1984; 1989a; 1989b; 1990a; 1993). This volume will provide a brief overview of these data, as well as a short discussion of some of the other Early Saxon faunal collections that have been excavated and analysed since the publication of the West Stow data.

The West Stow excavations yielded approximately 175,000 animal bones and fragments that could be reliably dated to the Early Saxon period. Detailed zooarchaeological analysis focused on the fauna that could be recovered from the SFBs, since these animal bone remains could be most closely dated.

Quantitative analyses indicated that the West Stow faunal collection was dominated by the remains of domestic mammals. Sheep were the most common species on the basis of NISP, followed by cattle, pigs, and a small number of horses. The presence of butchery marks on the horse bones indicates that they formed an occasional part of the Early Saxon diet. The species ratios based on NISP are shown in Figure 2.3. The domestic mammals were supplemented by domestic fowl (*Gallus gallus*) and geese (*Anser anser*), as well as smaller numbers of wild animals including red deer (*Cervus elaphus*), roe deer (*Capreolus capreolus*), and water birds, such as swans (*Cygnus* sp.) and waders, such as cranes (*Grus grus*).

The age distributions for the main domestic mammals suggest a lack of economic specialisation. A substantial number of sheep were killed during the first two years of life, which is consistent with a milk- or meat-production strategy. However, the presence of a number of older animals, combined with the evidence for loom weights

Site Name	Date	Citation	%COW	%S/G	%PIG	NISP
West Stow	ES 5th century	Crabtree (1990a)	33	45	22	7701
West Stow	ES 6th century	Crabtree (1990a)	35	51	14	13667
West Stow	ES late 6th–7th century	Crabtree (1990a)	34	47	20	1556
West Stow Visitor Centre	ES ?6th century	Crabtree (unpublished)	43	51	7	2146
Kilham	ES	Archer (2003)	46	49	5	2579
Spong Hill	ES	Bond (1995)	84	12	4	588
Melford Meadows	ES	Powell and Clark (2002)	61	29	10	479
Quarrington	ES	Rackham (2003)	63	25	11	1004
Redcastle Furze	ES 6th–7th century	Wilson (1995)	51	35	14	1201
Gamlingay	ES	Roberts (2005)	44	37	20	1032

Table 2.1 Species ratios (based on NISP) for Early Saxon sites in eastern England

from three SFBs, suggest that some small-scale wool production may have taken place as well. Similarly, the ageing evidence for cattle provides no evidence for a specialised meat-producing economy. The ageing data, when combined with the evidence for traction pathologies (Crabtree 1990a, 75) suggest that the West Stow Anglo-Saxon cattle were used for meat, milk, and traction.

As noted above, excavations were recently carried out at West Stow in advance of the construction of a new visitors' centre. The excavations revealed six additional structures, including five SFBs, and produced about 7000 additional animal bones. From a landscape perspective, these new data suggest that West Stow was a sprawling Anglo-Saxon settlement, similar to the Early Saxon settlement of Mucking in Essex (Hamerow 1993), rather than a small, bounded village. The new faunal data are generally similar to the original West Stow material, with a few exceptions. Species ratios based on NISP suggest that sheep were only slightly more numerous than cattle in the new assemblage (Figure 2.4). In comparison with the original West Stow data, there are far fewer pigs in the new faunal collection. Ageing data for sheep (presented in Chapter 4) indicate that most of the sheep were slaughtered during the first two years of life, indicating that West Stow sheep husbandry was probably focused on meat, milk, and herd security, rather than on wool production.

In general, the West Stow Early Saxon economic pattern appears to be one of relative self-sufficiency. When the West Stow fauna were first studied in the late 1970s, there were almost no comparable Early Saxon faunal assemblages. However, the recent analysis of the faunal remains from the Early Saxon village at Kilham in East Yorkshire (Archer 2003) has provided an important comparandum. Like West Stow, the faunal assemblage from Kilham was dominated by the remains of domesticated sheep and cattle. Based on the kill-patterns, Archer (2003, 53) suggests that the Kilham cattle were kept for meat, milk, and traction and that the sheep were raised for their meat and their wool. The data from West Stow and Kilham both point to a degree of economic self-sufficiency during the Early Saxon period.

The West Stow and Kilham cattle were also quite similar in size. The Kilham cattle had an average withers height of 113.0cm (Archer 2003, 73). At West Stow, the 5th-century cattle had an average withers height of 111.7cm, and the 6th-century cattle have an average withers height of 114.0cm (Crabtree 1990a, 36–38). Both the Kilham and the West Stow assemblages lack the large

cattle, with withers heights of around 130cm, seen at sites such as Icklingham and Colchester.

The West Stow sheep, on the other hand, are somewhat larger than the Kilham sheep. The sheep from all three phases at West Stow have an average estimated withers height of about 62cm (Crabtree 1990a, 49) and those from the new Visitor Centre excavations have an average estimated withers height of 60.7cm, while the sheep from Kilham have an average withers height of only about 56cm (Archer 2003, table 13). The West Stow sheep appear to have maintained much of the size improvement that was introduced by the Romans.

Finally, the butchery marks on the cattle, sheep, pig and horse bones from West Stow suggest that the animals were butchered on an ad-hoc basis by individual farmers. They lack the systematic pattern of butchery seen on the domestic animals from Icklingham and other Roman sites, such as Winchester (Maltby 2010).

During the past 15 years, a number of smaller Early Saxon faunal assemblages have been analysed from a range of sites in eastern England, including Melford Meadows (Brettenham) in Norfolk (Powell and Clark 2002), Spong Hill in Norfolk (Bond 1995), Station Road, Gamlingay in Cambridgeshire (Roberts 2005), Redcastle Furze (Thetford) in Norfolk (Wilson 1995), and Quarrington in Lincolnshire (Rackham 2003).

The species ratios based on NISP for these Early Saxon sites have been included in Table 2.1, along with data from the original and new West Stow excavations and the Kilham project. Unlike West Stow and Kilham, the other Early Saxon sites are dominated by the remains of cattle. Cattle are particularly numerous at Spong Hill (Bond 1995), but the assemblage is relatively small and poorly preserved. However, cattle and sheep appear to be the predominant species in all cases. Pigs generally make up about 20% or less of the large domestic mammals.

The ageing data from Spong Hill and Melford Meadows are very limited. However, the data from the other sites suggest a pattern of animal use similar to what is seen at West Stow and Kilham. At Station Road, Gamlingay (Roberts 2005) the cattle shows peaks of slaughter at both the immature and the mature adult stages. In addition to meat, cattle provided milk, calves, manure, and traction. The sheep kill-pattern is similar to the West Stow mortality profile, with peaks at both the immature and the mature adult stages. At Redcastle Furze, younger sheep were slaughtered for their meat, while older animals were kept for their wool. At both Redcastle Furze and Quarrington (Rackham 2003), there is a concentration on younger cattle, probably for meat. The ageing data for sheep from Quarrington suggest 'a non-focused management strategy of a largely subsistence character' (Rackham 2003, 271). In summary, the zooarchaeological data from Early Saxon England suggest that the goal of animal husbandry practices was autarky or economic self-sufficiency (Crabtree 2010b).

V. Summary

While farmsteads, such as Iron Age West Stow, may have produced animal products designed primarily to meet local needs, other sites, such as Haddenham in the fenland, suggest that some Iron Age farmers in East Anglia were also engaged in more specialised trade and exchange in animal products. During the Roman period, both small towns such as Icklingham and larger urban centres such as Colchester were part of a more complex system of animal production and exchange. Beef was butchered and distributed by specialist butchers, and there is clear evidence for the appearance of larger, improved cattle by the late Roman period.

While Early Saxon animals may have maintained some of the size improvement that was introduced by the Romans, the nature of Early Saxon animal husbandry was very different from its Roman and Iron Age predecessors. Early Saxon animal husbandry was non-specialised and designed to meet local needs.

Chapter 3. The Species Present and their Relative Abundances

I. Introduction

This chapter will explore the quantitative and qualitative evidence for animal husbandry and hunting practices at Middle Saxon Ipswich, Wicken Bonhunt and Brandon. The broad themes that will be addressed include the range of animal species represented, the relative importance of herding and hunting in these Middle Saxon assemblages, and the taphonomic factors that may have affected species frequencies and body part distributions at these three sites. The questions that will be addressed include:

- What is the role of cattle, sheep, and pig husbandry at these three Middle Saxon sites?

- What roles did hunting and wildfowling play in these Middle Saxon economies?

- What can body part distributions tell us about Middle Saxon animal use?

- What is the evidence for commensal species, such as domestic dogs and cats?

- How do these assemblages compare to faunal assemblages from Early Saxon and Roman sites in

East Anglia, and what does this tell us about changes in animal economy through time?

A brief assessment of the overall faunal assemblages from these three sites is provided in the following section.

II. The Middle Saxon Faunal Assemblages

As noted in Chapter 1, this study is based on over 10,000 animal bones and fragments that were recovered from Middle Saxon contexts in Ipswich between 1974 and 1988, over 100,000 animal bones and fragments that were recovered from Middle Saxon features at Wicken Bonhunt in the 1970s, and over 158,000 mammal and bird remains that were recovered from Brandon between 1980 and 1989. A complete list of the mammal species identified at these three sites is included in Table 3.1.

Table 3.1 clearly shows that all three assemblages are dominated by the remains of domestic animals. The vast majority of the identified mammal bones are those of cattle, pigs, and caprines (sheep and/or goats). Hunting appears to have played a relatively minor role in the Middle Saxon economy in East Anglia. A closer

	Brandon	Ipswich	Wicken Bonhunt
Domestic Mammals			
Cattle (*Bos taurus*)	13441	4282	5138
Sheep (*Ovis aries*)	4799	46	555
Goat (*Capra hircus*)	21	40	5
Sheep/goat	19832	2120	3298
Pig (*Sus scrofa*)	9192	3130	20954
Horse (*Equus caballus*)	702	62	163
Dog (*Canis familiaris*)	151	17	26
Cat (*Felis catus*)	16	62	102
Wild Mammals			
Dolphin/small whale (cf. *Delphinus delphis*)	1		
Grey seal (*Halichoerus grypus*)	1		
Badger (*Meles meles*)	4		
Otter (*Lutra lutra*)	3	1	
Fox (*Vulpes vulpes*)			4
Hare (*Lepus* sp.)	28	1	4
Rabbit (*Oryctolagus cunniculus*)*	32		
Red deer (*Cervus elaphus*)	50	27	16
Roe deer (*Capreolus capreolus*)	108	12	143
Small mammal	7		
Small Artiodactyl	11488		25054
Large Artiodactyl	138		
Large mammal	8025		15517
Unidentified	86648		25775
Total	**154616**	**9800**	**96754**

*intrusive/not Anglo-Saxon

Table 3.1 Mammal species identified from Brandon, Ipswich and Wicken Bonhunt (NISP)

examination of the domestic mammal bones identified from these three Middle Saxon assemblages can shed light on the patterns of domestic animal production and consumption in East Anglia during the Middle Saxon period.

III. The Domestic Mammals

Species Ratios

While domestic mammals are clearly a major component of the Middle Saxon faunal assemblages at Ipswich, Wicken Bonhunt and Brandon, the relative importance of these three species varies dramatically. There are many different ways to estimate the relative abundance of animal species in a faunal assemblage (see Lyman 2008 for an up-to-date review of this issue), but the two most commonly used methods are fragment counts (or NISP, Number of Identified Specimens Per taxon) and minimum number of individuals (or MNI). Lyman (2008, 81) has argued that, 'NISP is to be preferred over MNI as the quantitative unit used to measure taxonomic abundance'. McCormick and Murray (2007, 10) prefer the MNI, arguing that, 'Unless one can assume equal degrees of butchery, similar taphonomy and consistent retrieval methods, one cannot reliably use fragment values [NISP] as a dependable method for inter-site comparisons'. Here, the species ratios will be based on NISP, since, as Lyman (2008, 79) has argued, 'NISP is more fundamental, less derived, and the two [MNI and NISP] generally provide redundant information'. Body-part frequencies will then be examined for the domestic mammal species in order to study the taphonomic history of these three faunal collections.

The species ratios for the large domestic mammals (cattle, sheep/goat, pig, and horse) based on NISP are shown in Table 3.2 and Figure 3.1. These data clearly show that the Brandon assemblage is dominated by caprines and that the Wicken Bonhunt assemblage is dominated by pigs. Cattle are the most common species at Ipswich, followed by pigs. Horse bones are poorly represented in all three assemblages.

The differences in species ratios seen at these three Middle Saxon sites are mirrored at other Anglo-Saxon sites in eastern England. The faunal assemblage from

	Cattle	Sheep/goat	Pig	Horse
Brandon NISP	13441	24652	9121	702
Brandon %NISP	28.1	51.4	19.0	1.5
Ipswich NISP	4282	2206	3130	62
Ipswich %NISP	44.2	22.8	32.3	0.6
W. Bonhunt NISP	5138	3858	20954	163
W. Bonhunt %NISP	17.1	12.8	69.6	0.5

Table 3.2 Species ratios for the large domestic mammals from Brandon, Wicken Bonhunt and Ipswich, based on NISP

Middle Saxon Ipswich is dominated by the remains of cattle, as are the assemblages from the other *emporia*, including Anglian York, Hamwic, and Anglo-Saxon London. At the Fishergate site in York, the 'Anglian samples were characterised by a very low diversity of taxa, with cattle clearly predominant' (O'Connor 1991, 294). At Anglo-Saxon Southampton, cattle make up 52% of the assemblage by fragment count, and 75% by weight (Bourdillon 1988, 181). Cattle were also the dominant species recovered from the recent excavation of the Middle Saxon site at the Royal Opera House in London (Rielly 2003, 318). A number of these deposits include large numbers of head and foot bones of cattle, suggesting that they represent primary butchery waste (Rielly 2003, 318–9). This, in turn, indicates that specialised butchers were present in London during the Middle Saxon period. As noted in Chapter 2, specialised butchers have been identified at Roman towns like Icklingham, Winchester (Maltby 2010) and possibly Colchester (Luff 1993), but they are absent from Early Saxon sites.

The rural Saxon sites are more varied. Both the Early Saxon sites of West Stow in Suffolk (Crabtree 1989a; 1990a) and Kilham in Yorkshire (Archer 2003) are dominated by sheep, with cattle as a strong second. A similar pattern is seen at Early Saxon West Heslerton in Yorkshire (Dobney *et al.* 2007, 223). Most other Early Saxon sites are dominated by the remains of cattle, with sheep taking second place (see Table 2.1). Sheep are also

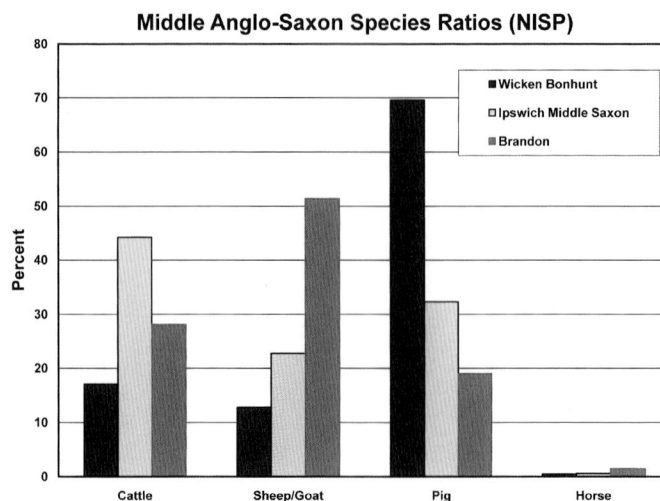

Figure 3.1 Species ratios for the large domestic mammals from Brandon, Wicken Bonhunt and Ipswich, based on NISP

Site Name	Date	Citation	%COW	%S/G	%PIG
West Heslerton	ES	Dobney *et al.* (2007, 223)	44	47	9
Station Rd, Gamlingay	ES	Roberts (2005)	44	37	20
Oakley Rd, Clapham	ES–MS	Maltby (nd a)	51	24	25
Mucking	ES	Done (1993)	63	16	21
Oxford Science Park	ES 6th–7th century	Ingram (2001)	59	24	17
Wicken Bonhunt	Late ES late 6th–7th century	Crabtree and Stevens (1995)	21	17	62
Harrold	ES–MS (400–850)	Maltby (nd b)	53	28	19
St Alban's Abbey	ES/MS 5th–8th/9th century	Crabtree (nd)	19	11	70
Bloodmoor Hill	Early MS	Higbee (2009)	60	19	21
St Alban's Abbey	MS 8th–9th century	Crabtree (nd)	15	14	71
Brandon	MS	Crabtree and Campana (2012b)	29	52	19
Quarrington	MS	Rackham (2003)	59	35	6
Fenland Management Proj.	MS	Baker (2002)	45	50	5
Flixborough	MS Phase 2–3a	Dobney *et al.* (2007, 223)	40	33	27
Flixborough	MS Phase 3b	Dobney *et al.* (2007, 223)	43	33	24
Flixborough	MS Phase 4–5b	Dobney *et al.* (2007, 223)	30	40	30
West Heslerton	MS	Dobney *et al.* (2007, 223)	39	53	8
Higham Ferrers	MS mid–late 8th century	Evans (2007)	33	34	33
Higham Ferrers	MS late 8th–early 9th century	Evans (2007)	57	25	19
Wicken Bonhunt	MS	Crabtree and Stevens (1995)	17	13	70
West Fen Rd, Cambs.	MS	Higbee (2005)	46	46	8
Yarnton/Cresswell/Worton	MS	Mulville and Ayres (2004)	56	30	14
Eynsham	MS Phase 2b	Mulville (2003)	18	61	21

Table 3.3 Summary of species ratios for cattle, sheep/goat, and pig for non-urban Early and Middle Saxon sites not included in Table 2.1

the most common species at the Middle Saxon site of Brandon and at the Phase 4–5b contexts at Flixborough in Lincolnshire (Dobney *et al.* 2007, 223). The earlier Middle Saxon contexts at Flixborough (Phases 2–3a and 3b), on the other hand, are dominated by cattle, followed by sheep (Dobney *et al.* 2007). The high proportion of pigs seen at Wicken Bonhunt is almost unique in the Anglo-Saxon archaeological record. The only comparable assemblage is the unpublished assemblage from the 5th- to 9th-century Chapter House at St Albans Abbey (Crabtree nd). These data are summarised in Figure 3.2 and Tables 2.1 and 3.3. The figure highlights the unusual character of both the Wicken Bonhunt and the St Albans Abbey assemblages.

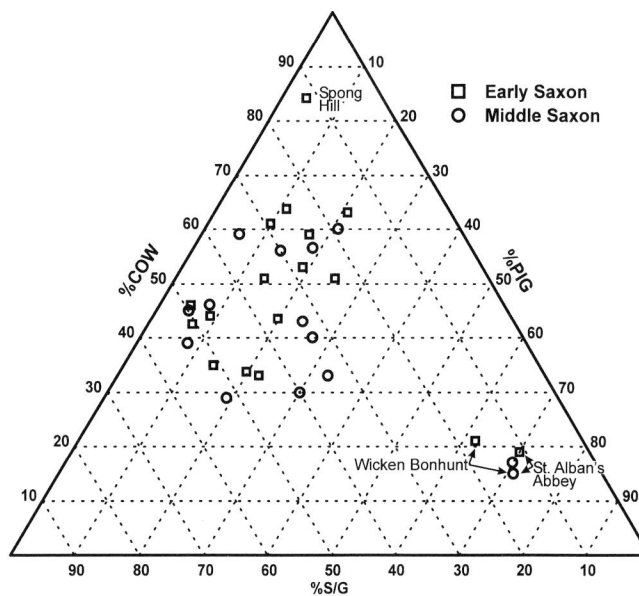

Figure 3.2 Species ratios based on NISP for cattle, sheep and pigs from Early and Middle Saxon rural and ecclesiastical sites in eastern England

	Cattle	Sheep/goat	Pig	Horse
Skull	314	123	6405	5
Horn core	327	124		
Maxilla	164	48	1640	0
Mandible	697	425	2925	6
Atlas	8	12	35	0
Axis	11	10	18	1
Sacrum	4	1	0	1
Vertebrae	2	4	78	1
Ribs	0	0	1	0
Innominate	137	151	182	12
Femur	158	55	107	19
Patella	3	0	0	2
Tibia	217	315	170	11
Fibula			135	0
Scapula	147	224	342	4
Humerus	142	121	206	6
Radius	158	174	126	18
Ulna	104	65	172	10
Astragalus	43	6	10	6
Calcaneus	103	19	42	4
Centro-quartal	7	1		
Tarsals	1	0	0	2
Carpals	20	0	0	2
Metatarsus	101	70	269	1
Metacarpus	74	41	250	4
Metapodium	15	7	76	1
1st phalanx	46	6	29	2
2nd phalanx	9	0	3	1
3rd phalanx	19	0	8	0
Tooth fragments	36	26	796	0
Loose teeth	724	349	3126	22

Table 3.4 Body part distribution for the large domestic mammals from Wicken Bonhunt

Body-part distributions

The analysis of body-part distributions is crucial to our understanding of the natural and cultural processes that have affected these faunal assemblages. Bone preservation can be affected by the method of disposal and rapidity of burial, the degree of carnivore activity, soil pH and the depositional environment, as well as the use of screening during excavation. The body-part distributions for Wicken Bonhunt (Table 3.4), Ipswich (Table 3.5), and Brandon (Table 3.6) have been presented in tabular form. Several conclusions can be drawn from these basic data. The first is that nearly all body parts for all four main large domestic species are present in all three assemblages. The second is that there are real variations in the frequencies of different elements both within the three assemblages and also between the three assemblages.

The presence of nearly all body parts for all the large mammal species suggests that whole animals were butchered and consumed on all three sites on a fairly regular basis. This is not a surprising conclusion for the rural sites of Brandon and Wicken Bonhunt, since it is certainly reasonable to conclude that at least some of the animals that were raised on these sites were slaughtered for home consumption. Ipswich, on the other hand, is an early urban site, and the body-part data suggest that many of the animals that were consumed at Ipswich were driven to the site on the hoof and then butchered in town. It is also possible that some of the animals consumed in Ipswich were raised on the outskirts of the town. A possible farm has been identified at the National Gallery Site on the outskirts of Lundenwic (Blackmore 2002).

One obvious characteristic of the body-part representation is that second phalanges of cattle, sheep, and pigs are consistently less numerous than first phalanges. For example, the Middle Saxon contexts from Brandon yielded 292 first phalanges of sheep and goats and only 78 caprine second phalanges. Sheep and goats have the same numbers of first and second phalanges, but the second phalanx is about half the size of the first phalanx. Maltby (2002) has suggested that the ratio of second to first phalanges can be used as a measure of bone recovery. In the case of these three sites, the bones were hand-collected without fine screening, and this is certainly responsible for the low numbers of carpals, tarsals, second phalanges, and sesamoids. At Brandon, a portion of the site was subject to fine screening. Examination of the screened samples revealed higher numbers of lateral pig phalanges and other small elements that are likely to be missed without fine sieving.

Zooarchaeologists and palaeontologists have often used the MNE (Minimum Number of Elements) as a way of examining more complex patterns of body part

	Cattle	Sheep/goat	Pig	Horse
Skull	238	171	347	0
Horn core	163	74		
Maxilla	32	20	102	1
Mandible	333	316	393	1
Hyoid	6	1	0	0
Atlas	15	3	4	0
Axis	6	5	2	0
Innominate	91	82	69	1
Femur	52	41	84	1
Patella	3	1	0	1
Tibia	200	192	153	2
Fibula			44	0
Scapula	173	230	259	3
Humerus	206	224	195	0
Radius	225	151	177	4
Ulna	85	43	157	4
Astragalus	191	35	51	4
Calcaneus	171	50	121	2
Tarsals	74	5	8	2
Carpals	78	3	1	0
Metatarsus	176	85	137	5
Metacarpus	175	109	166	2
Metapodium	137	51	39	3
1st phalanx	466	55	86	5
2nd phalanx	218	4	33	5
3rd phalanx	224	3	10	6
Tooth fragments	12	0	58	0
Loose teeth	446	204	227	9

Table 3.5 Body part distribution for large domestic mammals from Ipswich

representation. The MNE has been defined as 'the minimum number of skeletal elements necessary to account for an assemblage of specimens of a particular skeletal element' (Lyman 1994: 289). In simpler terms, the analyst is calculating the MNI for each skeletal element. These measures are often corrected or normed to reflect the number of times a particular element occurs in the skeleton. For example, a cow has 8 first phalanges, but only two humeri. These corrected measures are referred to as MAU or Minimal Animal Units, following Binford (1984, 50). The most common element can be set at 100%, and the other elements can be expressed as a percentage of the most common element.

This method was used to examine the patterns of body part representation for pigs at Wicken Bonhunt, Ipswich, and Brandon. As can be seen in Tables 3.4–3.6, mandibles are the most common pig bone elements represented in the assemblages from all three sites. The MNEs for scapulae, humeri, radii, metacarpi, femora, tibiae, and metatarsi were calculated and normed to reflect the fact that a pig has four major metacarpi but only two humeri. Since pig mandibles were the most common elements in all three assemblages, the other elements are expressed as percentages of the frequency of the mandibles. The results are shown in Figure 3.3. The figure shows that pig mandibles are far more common than all other skeletal elements at Wicken Bonhunt. They are ten times as common as almost any other skeletal element. At Ipswich

and Brandon, the post-cranial skeletal elements of pigs are much more common. The Ipswich assemblage is similar to many other medieval urban assemblages where pig cranial elements outnumber postcranial remains. However, the imbalance between cranial and postcranial remains is far more marked at Wicken Bonhunt.

The results for cattle bones are shown in Figure 3.4. While mandibles outnumber post-cranial elements at all three sites, the proportions of the elements are much less skewed. At Ipswich and Wicken Bonhunt, the post cranial elements are about 60% as common as the mandibles. Post-cranial elements are relatively less common at Wicken Bonhunt, but the cattle limb bones are more than twice as common as the pig post-cranial elements. The sheep and goat data (Figure 3.5) are very similar for all three sites. Mandibles and tibiae are generally well represented, and other post-cranial elements are somewhat less common.

There are several possible reasons why pig postcranial elements are so poorly represented at Wicken Bonhunt. One possibility is that this disparity reflects recovery or identification bias. Since these three assemblages were recovered and analysed in similar ways, the pattern seen at Wicken Bonhunt is unlikely to be the result of inconsistencies in recovery or identification. The pattern may reflect the destruction of less dense, unfused postcranial elements by dogs and other agents of bone destruction. If this were the explanation for the body-part

The data from the Middle Saxon sites in Ipswich present a very different picture. Here, Pat Stevens distinguished sheep from goats on the basis of their horn cores. She identified 46 sheep and 40 goat horncores from Middle Saxon contexts at Ipswich. These goat horns may have provided the raw material for horn-working. At Hamwic, Jennifer Bourdillon (1988, 182) noted that goat bones were quite rare and that most of the identified goat remains were horn cores. She stated that:

> The few postcranial goat bones have been butchered and would represent food waste, but the animals were clearly not reared for the special purpose of being sent to town for food. There is no evidence of the slaughter of kids for skin, and it seems that the main importance of goats for Hamwic would have lain in the industrial raw material of their horn.

At the *emporium* of Dorestad in the Netherlands, most of the identifiable goat remains were horn cores, and most of the horn cores came from male animals. Prummel (1983, 196) suggested that goat horns and skins were imported into Dorestad for industrial purposes, and that few goats were kept at the site.

Horses

Although horse remains are rare at Anglo-Saxon sites in England, butchery marks on the horse bones from Early Saxon West Stow indicate that horses were used for food, at least on an occasional basis, in Early Saxon times. The crucial question is whether horses continued to form a part of the Middle Saxon diet, since Pope Gregory III (*c*. 732 AD) prohibited the consumption of horseflesh by Christians in a letter sent to St Boniface, apostle to the Germans. The extent to which this prohibition was enforced in 8th- and 9th-century England remains an open question.

Horses make up less that 1% of the large domestic mammal remains at Wicken Bonhunt. They appear to have been used for food, at least on an occasional basis, since 59 of the horse limb bones showed butchery traces. The butchery marks appear on 6 humeri, 10 ulnae, 4 metacarpi, 14 pelves, 19 femora, and 6 metatarsi. The butchery includes several examples of splitting the long bones, presumably for marrow extraction. There is no pathological evidence to suggest that the Wicken Bonhunt horses were used for traction, but four fused thoracic vertebrae may indicate the presence of spondylitis which can be caused by breaking a horse for riding.

The data from Middle Saxon Ipswich also suggest that horseflesh was eaten, at least on an occasional basis. Although horses made up less than 1% of the large domestic mammal assemblage, chop marks were apparent on some limb bones, and one metcarpus was split for marrow removal.

Hippophagy is documented at other Middle and Late Saxon sites in Eastern England. At Flixborough in Lincolnshire, there is evidence for butchery on horse bones from all Anglo-Saxon phases (Dobney *et al.* 2007, 111). The evidence for horse consumption at Hamwic is less clear. Bourdillon (1988, 182) notes that horses were eaten seldom, if ever, in Anglo-Saxon Southampton. Their bones are rarely found in pits with other food waste, and many of the Hamwic horses appear, on the basis of dental

wear, to have been very old. The Hamwic horses may have been used for riding or carrying packs.

The evidence from Brandon presents a more complex picture. Horses make up 1.5% of the large domestic mammal remains on the basis of NISP (and about 1.4% on the basis of MNI). The Brandon assemblage included the remains of both juvenile and senile horses, so it is possible that horses were bred at the site.

While horse bones made up a small percentage of the Brandon faunal assemblage, horses do appear to have played a ritual role at the site. A horse skull, fragmentary mandible, and portions of the axial skeleton were found in the door pit leading to the chancel of the timber building that has been identified as a church. The horse was an elderly male animal with very heavily worn teeth. Two of the vertebrae showed slight lipping and fusion. The horse may be a foundation deposit.

Although the use of horses as foundation deposits is rare in Anglo-Saxon archaeology, horse burials are common from ritual contexts in Early Saxon East Anglia. For example, the burial of a horse head was recovered from a wealthy Early Saxon burial at the Snape cemetery, and horse burials have also been recovered from Sutton Hoo and Eriswell in Suffolk (Filmer-Sankey and Pestell 2001, 256). The horse head burial at Snape shows some interesting similarities to the Brandon foundation deposit. The Snape horse, like the Brandon deposit, was an elderly male animal. Simon Davis (2001) noted that while crown height estimates for the Snape horse teeth indicated that the stallion was more than seventeen years old, the degree of wear on the teeth suggested that the animal was between 20 and 30 years old at the time of its death.

Chris Fern (2007) has recently summarised the evidence for Early Saxon horse burials between the 5th and the 7th century. Horse cremations are common in the regions of the Humber Estuary, the Wash, and northern Norfolk. The cemetery of Spong Hill in Norfolk alone included 227 horse cremations among the more than 2000 cremation burials that were excavated at the site (see Bond 1994). Horse cremations in eastern England are not sex-specific. They are found with both men and women, and occasionally with children (Fern 2007, 99). Horse burials, on the other hand, are associated exclusively with male elites. Many of the known horse burials are concentrated in the Lark and Cam River Valleys in East Anglia. Fern (2007, 102) has suggested that they may represent competition between elite groups. It is possible that the Brandon foundation deposit reflects elite status in an otherwise ostensibly Christian context.

Commensal Species

Ipswich, Wicken Bonhunt and Brandon all yielded the remains of commensal mammals, although the numbers of cats and dogs vary from site to site. Brandon produced 151 dog bones and only 15 domestic cat bones, while cat bones were far more common than dog bones at both Ipswich and Wicken Bonhunt. Dog bones were well represented at the Early Saxon site of West Stow, including two complete dog skeletons that were recovered from SFB 16 (Crabtree 1990a, 62–67). Early medieval historical sources, such as the 10th-century Laws of Hywel Dda of Wales (Clutton-Brock 1976, 385) and Aelfric's Colloquy (Watkins 2010), indicate that dogs were used both for hunting and for herding in early medieval Britain.

	West Stow	Brandon	Wicken Bonhunt	Ipswich
Domestic Birds				
Chicken (*Gallus gallus*)	521	1306	3082	513
Goose (*Anser anser*)	285	964	2038	126
Duck/mallard (*Anas platyrhynchos*)	27	755	147	9
Duck/goose (Anatidae)			2	
Peafowl (*Pavo cristatus*)			2	
Doves cf. domestic dove (*Columba livia*)			3	

Table 3.7 Domestic bird species identified from Brandon, Ipswich, Wicken Bonhunt and West Stow

Anglo-Saxon cats were certainly valuable as mousers, but whether they were feral barn cats, pampered pets, or a source of skins remains a matter of debate (O'Connor 1992). The Early Saxon site of West Stow produced four cat bones with skinning marks, suggesting that at least some Anglo-Saxon cats were used for their fur (Crabtree 1990a, 104–5). No such marks were seen on the cat bones from the Middle Saxon sites of Brandon, Ipswich and Wicken Bonhunt. The substantial number of cat bones recovered from the Middle Saxon contexts at Wicken Bonhunt (102 bones) and Ipswich (62 bones) may reflect their importance in eradicating small vermin.

IV. Domestic Birds

The domestic bird remains recovered from Brandon, Wicken Bonhunt, and the Middle Saxon contexts from Ipswich are shown in Table 3.7. The data from the original excavations at West Stow were included for comparison. While the rural sites of Wicken Bonhunt, Brandon, and West Stow produced a diverse range of wild bird remains (see below), the vast majority of bird remains from all three Middle Saxon sites and West Stow are domestic species, primarily domestic chickens (*Gallus gallus*) and domestic geese (*Anser anser*). All four sites also produced the remains of ducks (*Anas platyrhyncos*), although it is not possible to determine whether these are domestic ducks or mallards. Domestic birds would have provided the Anglo-Saxons with a range of primary and secondary products, including eggs, meat and feathers. In addition, geese are excellent 'watchdogs'; they will honk loudly when anyone approaches them.

To assess the relative importance of domestic birds in the Middle Saxon economy, the relative importance of cattle, caprines, pigs, chickens and geese was calculated based on NISP (*cf.* Dobney *et al.* 2007, 117–119). The results are shown in Table 3.8, along with the data from the Early Saxon SFBs at West Stow, Suffolk, and the data from the Early Saxon site of Kilham in East Yorkshire (Archer 2003). These data allow us to make several observations. First, at all the East Anglian sites, chicken bones outnumber goose remains. At Kilham, in contrast, goose bones are more numerous than chicken remains. The relative importance of bird remains is highest at the rural site of Wicken Bonhunt, where bird bones make up nearly 15% of the five major domestic animals. These data are similar to the results from the Middle and Late Saxon site of Flixborough in Lincolnshire, where domestic birds make up between 10 and 33% of the major domestic animals (Dobney *et al.* 2007, 117). Domestic birds appear to play a smaller role at the *emporium* of Ipswich and at the rural site of Brandon. However, the proportions of domestic birds at Ipswich and Brandon are higher than they are at the Early Saxon site of West Stow.

V. Wild Mammals

While the bones of wild mammals are present at all three Middle Saxon sites, hunting appears to have played a relatively minor role in the Middle Saxon economy of East Anglia. One way of assessing the importance of hunting is to compare the number of wild mammal bones as a percentage of all identified mammal bones, excluding the commensal species. Using this measure, wild mammals make up 0.4% of the Brandon assemblage, 0.42% of the Ipswich assemblage, and 0.55% of the Wicken Bonhunt mammal bone assemblage. Roe deer are the most common hunted species at both Wicken Bonhunt and Brandon, while red deer bones are about twice as common as roe deer at Ipswich.

	Cattle	S/G	Pig	Chicken	Goose	**Total**
West Stow	7873	11148	3803	521	285	**23630**
Kilham	1199	1254	126	37	88	**2704**
Brandon	13441	24652	9121	1306	964	**49484**
Wicken Bonhunt	5138	3858	20954	3082	2038	**35070**
Ipswich	4282	2206	3130	513	126	**10257**
	% Cattle	%S/G	%Pig	%Chicken	%Goose	
West Stow SFBs	33.3	47.2	16.1	2.2	1.2	
Kilham	44.3	46.4	4.7	1.4	3.3	
Brandon	27.1	49.8	18.4	2.6	1.9	
Wicken Bonhunt	14.6	11.0	59.7	8.8	5.8	
Ipswich	41.7	21.5	30.5	5.0	1.2	

Table 3.8 Relative importance (based on NISP) of cattle, sheep/goat, pigs, chickens and geese from the Early Saxon sites of Kilham and West Stow and the Middle Saxon sites of Brandon, Ipswich and Wicken Bonhunt

Plate 3.1 Male red deer skull from Brandon. The antlers have been removed by chopping and sawing

While Brandon had the smallest percentage of wild mammals, it also produced the most diverse wild mammal assemblage. In addition to the remains of red and roe deer, small numbers of hare and fox bones were recovered from Wicken Bonhunt, and the Middle Saxon assemblages from Ipswich yielded a single bone each of hare and otter. Brandon, however, produced small numbers of hare, otter, and badger bones, plus single bones of grey seal and dolphin/small whale. Cetacean bones are not common on medieval archaeological sites in Britain (Gardiner 1997, 189). Large numbers of cetaceans were recovered from the Middle and Late Saxon site of Flixborough in Lincolnshire (Dobney *et al.* 2007, 48–51), and the analysts have argued that the consumption of cetaceans may well be associated with high status individuals in the Middle and Late Saxon periods. If this is the case, then the presence of cetacean bone at Brandon may be one more indicator that Brandon was a high-status Middle Saxon site.

Wild mammal bones and deer antlers also played a role in Anglo-Saxon crafts. The crafts of bone- and leather-working were practiced at Ipswich (Wade 2000), and the zooarchaeological evidence from Brandon suggests that both antler and bone may have served as raw materials for craft production. At Brandon, a large male deer skull, Plate 3.1, had one antler removed by sawing, and the other chopped off. While there is no direct evidence for bone-working at Brandon, a single context, 4947, yielded 18 red deer metapodia. These bones may well represent a cache of bones that were set aside as raw materials for bone artefact manufacture. The long, straight, thick-walled shafts of deer metacarpals and metatarsals make them ideal materials for bone working.

VI. Wild Birds

The rural Middle Saxon sites both produced rich wild bird assemblages; including many water birds and waders, species that would have been abundant in East Anglia. Table 3.9 lists the wild bird remains that were recovered from Brandon, Ipswich, and Wicken Bonhunt. The wild bird assemblage from the original West Stow excavations is included for comparative purposes. The Wicken Bonhunt assemblage produced the remains of several species of wild ducks, pink-footed goose (*Anser brachyrhynchos*), swan, bittern (*Botaurus stellaris*), and cranes (*Grus* sp.). The Brandon assemblage yielded the remains of wild ducks, a bittern, a diver (*Gavia stellata*), cranes and swans. The Early Saxon site of West Stow also yielded the remains of water birds and waders including white-fronted goose (*Anser albifrons*), wild ducks, grey heron (*Ardea cinerea*), swans and cranes. A crane bone was also recovered from the West Stow Visitor Centre site. These data suggest that fowling played a small but significant role in Early and Middle Saxon rural economy.

The remains of cranes are particularly interesting. They appear in substantial numbers at West Stow, Brandon and Wicken Bonhunt. The Brandon remains were an excellent match for sub-fossil (early Holocene) East Anglian crane remains housed in the Cambridge University Museum of Zoology. Cranes bred in East Anglia until about 1600. Common cranes (*Grus grus*) have recently been sighted in Norfolk and Suffolk, and they are now breeding in the RSPB reserve at Lakenheath, Suffolk for the first time in about 400 years (Royal Society for the Protection of Birds, 2007). The presence of crane remains in rural East Anglian sites suggests that they must have been quite widespread in the later first millennium. Remains of cranes were also recovered from 7th- to 10th-century contexts at Flixborough in Lincolnshire (Dobney *et al.* 2007, 48). Albarella and Thomas (2002) have argued that wild birds such as cranes would have played a minor role in the medieval diet, but they may have served as luxury foods. Their presence on high status sites, including Brandon, Wicken Bonhunt and Flixborough is notable in this regard. Sykes (2004, 98), however, notes that cranes are common at most types of archaeological sites prior to the 12th century and suggests that they are not necessarily marks of elite status.

The most striking feature of the wild bird remains from Middle Saxon East Anglia is the presence of a nearly complete peregrine falcon (*Falco peregrinus*) at Middle Saxon Brandon (Plate 3.2). This specimen is the earliest known archaeological example of a peregrine falcon from Anglo-Saxon England. Historians such as Vandervell and Coles (1980, 29) have assumed that the history of hawking in England dates back to the 7th or 8th century. Falcons are mentioned in 8th-century letters, and hawkers are recorded as members of the royal household of Mercia in the late 8th century (Dobney *et al.* 2007, 241–242). However, until the discovery of the Brandon falcon, the archaeological evidence for Middle Saxon falconry was lacking. The natural habitats of peregrine falcons are cliffs and upland areas (Lascelles 1892, 236), so the peregrine from Brandon is almost certainly a captive bird. However, peregrines do occasionally appear in East Anglia as non-breeding visitors (Heinzel *et al.* 1972).

On the basis of metrical data, the Brandon falcon is clearly a female. In Table 3.10, the measurements taken on

Wild Birds	West Stow	Brandon	Wicken Bonhunt	Ipswich
Diver (*Gavia stellata*)		2		
Duck (*Anas* sp.)	1	77	5	
Pink-footed goose (*Anser brachyrynchos*)			1	
White-fronted goose (*Anser albifrons*)	2			
Teal (*Anas crecca*)	3		2	
Garganey (*Anas querguedula*)			3	
Shoveller (*Anas clypeata*)			1	
Pochard (*Aythia ferina*)			2	
Pintail (*Anas acuta*)			2	
Wigeon (*Anas penelope*)			1	
Shelduck (*Tadorna tadorna*)			1	
cf. Smew (*Mergus albellus*)			2	
Coot (*Fulicia atra*)			1	
East Anglian crane (*Grus grus*)	30	19	26	
Grey heron (*Ardea conerea*)	1			
Peafowl (*Pavo cristatus*)			2	
Swan (*Cygnus* sp.)	2	11	2	1
Bittern (*Botaurus stellaris*)		1	1	
cf. Black grouse (*Lyrurus tetrix*)			1	
Wood pigeon (*Columba palumbus*)			44	
Stock dove (*Columba oenas*)			2	
Pigeon/dove (*Columba* sp.)		3	2	3
Plovers (*Pluvialis* sp.)	1		2	
Grey plover (*Pluvialis squatarola*)	5			
Lapwing (*Vanellus vanellus*)	1			
Woodcock (*Scolopax rustica*)	1			
Herring/Lesser black-backed gull (*Larus* sp.)	1			
Common gull (*Larus canus*)	2			
Snipe (*Gallinago gallinago*)	1			
Thrushes (Turdidae)	1	1		
Song thrush (*Turdus philomelos*)	7			
Starling (*Sternus vulgaris*)	2			
Crows (Corvidae)		10	2	2
Raven (*Corvus corax*)			8	1
Jackdaw (*Corvus monedula*)			1	3
Gannets (Sulidae)				2
Common buzzard (*Buteo buteo*)	1	1	7	
Peregrine falcon (*Falco peregrinus*)		18		

Table 3.9 Wild bird species identified from Brandon, Ipswich, Wicken Bonhunt and West Stow

the Brandon falcon are compared to measurements taken on male and female peregrine falcons by Solti (1985). Female birds were often preferred for falconry.

In addition, the falcon from Brandon was discovered as a nearly complete skeleton. The fact that the Brandon falcon was buried intact suggests that it was a valued possession. Falconry is traditionally a sport of the upper classes, and the presence of a nearly complete falcon at Brandon, when combined with the rich artefactual evidence from the site, supports the interpretation of Brandon as a high status site. Sykes (2004, 99) argues that while wild bird hunting was not necessarily an indicator of high social status during the Middle Saxon period, hawking was. Moreover, Dobney and Jacques (2002, 14) suggest that bitterns were taken primarily through falconry, and bitterns are present at both Brandon and Wicken Bonhunt. Falcons and hawks can be used to hunt cranes, which are also found at both sites.

Measurement (mm)	Brandon	Female Mean	Male Mean
Humerus (GL)	88.1	88.5	78.0
Femur (GL)	72.2	70.6	64.5
Tibiotarsus (GL)	91.0	92.9	81.5
Tarsometatarsus (GL)	53.8	54.9	49.1

Table 3.10 Measurements taken on the Brandon falcon compared to those from male and female peregrine falcons (Solti 1985)

Dobney *et al.* (2007, 243–44) have suggested that other raptors, such as the common buzzard (*Buteo buteo*) and red kite (*Milvus milvus*) may also have been used in Anglo-Saxon hawking. Buzzard bones were recovered from the Middle-to-Late Saxon site of Flixborough in Lincolnshire, as well as from Wicken Bonhunt, from

Plate 3.2 Peregrine falcon from Brandon

Brandon, and even from the Early Saxon site of West Stow. While buzzards certainly can be trained for hunting, the fact that the buzzard bones from the Early and Middle Saxon sites in East Anglia were found fragmented in midden deposits argues against their use in hawking.

The wild bird remains from the *emporium* of Ipswich present a very different picture. Aside from the domestic ducks/mallards, there are very few wild bird remains at Ipswich. The remains of water birds are limited to a single bone of a swan and two bones of gannet. These data suggest that bird hunting played a very minor role in Middle Saxon Ipswich. The most common wild species are corvids, including a single bone of a raven (*Corvus corax*), plus the remains of jackdaws (*Corvus monedula*) and crows (Corvidae). Several species of corvids were also discovered at the late Roman site of Icklingham including carrion crow (*Corvus corrone*), jackdaw, and two raven bones associated with a human burial (Crabtree 2010a). Corvids can thrive in urban contexts, since they can scavenge human rubbish (Marzluff and Angell 2005, 292). This may explain their presence at the Roman small town of Icklingham and the Middle Saxon contexts at Ipswich.

The impoverished nature of the Ipswich avifauna is mirrored at other *emporia* in Britain. The Peabody site in London was part of the 7th- to 9th-century trading settlement of Lundenwic. The evidence for hunting at Middle Saxon London is limited to a small number of wild birds and a few post-cranial remains of red and roe deer (B. West 1989, 152). Evidence for hunting at the Royal Opera House site in London is limited to a small number of bones of red and roe deer, plus single bones of hare and whale. Bird remains include only chickens, geese, ducks, and a single bone of dove (*Columba* sp.) (Rielly 2003, table 72). O'Connor (1991, 294) has argued that hunting was also unimportant at Anglian York. These data suggest that the inhabitants of the British *emporia* were not regularly augmenting their diets by fowling and hunting.

VII. Discussion

The archaeozoological data indicate that domestic mammals played a major role in the economies and diets of the Middle Saxon inhabitants of East Anglia. The relative importance of the major domestic species, however, varies from site to site. Sheep are dominant at Brandon, as they are at the nearby Early Saxon site of West Stow. The Wicken Bonhunt assemblage is made up of nearly 70% pig bones. Cattle are the most common species at Ipswich, a pattern that is mirrored at the other *emporia* in Britain. Although horses were occasionally eaten, they played a very minor role in the Middle Saxon diet. However, their use as foundation deposits in an ostensibly Christian context indicates that they clearly played an important symbolic role in Middle Saxon society. In addition to the domestic mammals, the Middle Saxons raised chickens and geese, which would have provided meat, eggs and feathers.

Small numbers of wild mammals and birds supplemented the domestic species at all the Middle Saxon sites in East Anglia. The most commonly hunted mammals were red and roe deer. A diverse range of water birds and waders were hunted at both Brandon and Wicken Bonhunt. The faunal remains from Ipswich, on the other hand, provided little evidence for fowling.

In summary, the archaeozoological data from Brandon, Ipswich and Wicken Bonhunt indicate that there was a substantial diversity in animal exploitation during the Middle Saxon period in East Anglia. However, species ratios tell only part of the story. In the following chapter, ageing data and other lines of zooarchaeological evidence will be used to explore the ways that the primary and secondary products of these animals were utilised by the Middle Saxon inhabitants of East Anglia.

Chapter 4. Animal Exploitation in Middle Saxon East Anglia

I. Introduction

This chapter will explore how the Middle Saxons made use of their domestic animals. What roles did primary and secondary animal products (Sherratt 1981; 1983) play in the East Anglian economy? Much of the analysis will be based on ageing data, examining the ages at which the animals were slaughtered. For example, sheep that are raised primarily for meat are often killed at about 2–3 years of age, around the time that they reach bodily maturity. Continuing to feed an animal after that time will not lead to an increased meat yield. Generally, shepherds focusing on meat production will maintain a small breeding stock composed primarily of females. Farmers who are interested in wool production, on the other hand, may keep many older sheep, since sheep will continue to produce wool throughout their lifetimes. Castrated males, wethers, are particularly good wool producers (see Payne 1973). Older female sheep might be expected in a milk-producing flock, and documentary sources, such as Aelfric's Colloquy (Watkins 2010, 4), indicate that sheep were regularly milked during the Anglo-Saxon period.

The analyses of ageing will be based on two types of data: dental eruption and wear (Payne 1973; Grant 1982) and epiphyseal fusion of the limb bones (Silver 1969). The principles behind these methods are relatively simple and are based on mammalian biology. All terrestrial mammals have two sets of teeth: a set of deciduous or milk teeth and a set of permanent teeth which replace the milk teeth in a set sequence for each species. This process allows us to estimate the age of juvenile animals quite closely. Once the permanent teeth have erupted, they continue to wear throughout the animal's lifetime. By examining the degree of wear on the permanent teeth, analysts can estimate the age of adult mammals, especially those with relatively high-crowned or hypsodont teeth. In juvenile animals, the shafts of long bones (diaphyses) are separated from the joint ends (epiphyses) by cartilaginous plates. When bone growth is completed, the plates ossify in a set sequence. By examining the state of fusion of each of the limb bone epiphyses, archaeozoologists can construct mortality profiles for the domestic (and wild) mammals. There are two main problems with this method, however. First is that epiphyseal fusion is completed around the time an animal reaches bodily maturity. The method cannot be used to distinguish adult animals from elderly ones. The second problem is a taphonomic one. As Brain's (1967) ethno-archaeological research in southern Africa demonstrated many years ago, the fragile unfused limb bones of juvenile animals are more subject to carnivore ravaging than the denser bones of adult animals are. Despite these limitations, the faunal evidence from Wicken Bonhunt will demonstrate that epiphyseal data, when combined with data on tooth eruption and wear, can produce a fuller picture of animal exploitation in Middle Saxon East Anglia.

In this chapter, the faunal data from Brandon, Wicken Bonhunt and Ipswich will be used to address the following questions:

- Were Middle Saxon sheep raised primarily for meat, milk, wool, or some combination of these products?
- Were cattle reared for meat, milk, and/or traction?
- What roles did pigs play in Middle Saxon animal husbandry?
- Is there evidence for economic specialisation at the rural Middle Saxon sites of Brandon and Wicken Bonhunt?
- Were the inhabitants of the *emporium* of Ipswich provisioned with meat from the countryside?
- Is there archaeological evidence for trade in animals and animal products in Middle Saxon East Anglia?

II. Mortality Profiles for the Main Domestic Species

In order to answer these questions, the ageing data for each of the major mammal domestic species — sheep, cattle, and pigs — will be presented. At the end of the chapter, these data will be used to address the questions of economic specialisation and urban provisioning in Middle Saxon East Anglia.

Pigs

Pigs are raised for primary products, mainly meat, although the hides and bristles are also useful. Pigs are prolific; they reproduce at young ages and produce large litters. Unlike cattle and sheep, pigs are easy to keep in urban environments. They were reared in some New York City neighbourhoods (see, for example, Milne and Crabtree 2001) throughout the 19th century. Analyses of the ages at which pigs were killed can tell about the nature of Middle Saxon pig husbandry. The questions that can be addressed include:

- Were pigs raised for local consumption at the rural Middle Saxon sites?
- Were pigs traded between sites?
- Were the inhabitants of Ipswich provisioned with pigs of specific ages?

Wicken Bonhunt is an unusual Anglo-Saxon site because nearly 70% of the large domestic faunal remains were identified as pigs. As noted in Chapter 3, most other Anglo-Saxon sites are dominated by the remains of cattle and sheep. Mortality profiles for the Wicken Bonhunt pigs were based on mandible wear stages (MWS) following Grant (1982). The distribution of mandible wear stages for the 313 complete pig mandibles from the large boundary ditch that produced about 75% of the Middle Saxon fauna from Wicken Bonhunt, is shown in Figure 4.1. Relatively

Mandible Wear Stages for Wicken Bonhunt Pigs

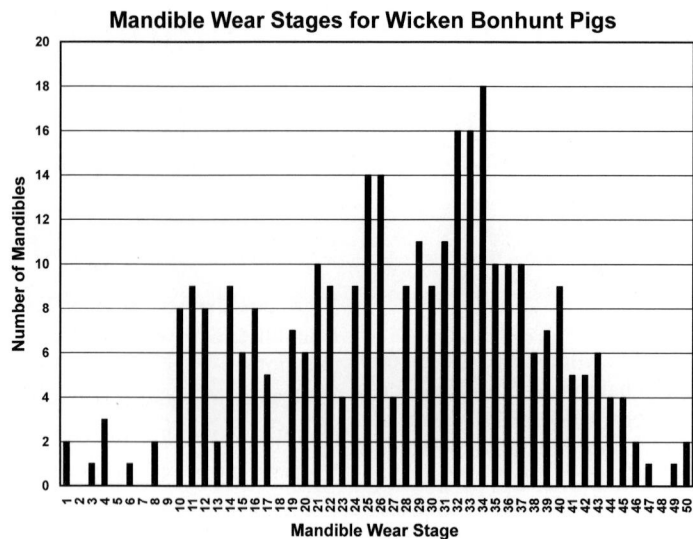

Figure 4.1 Distribution of Mandible Wear Stages for pig mandibles from Middle Saxon Wicken Bonhunt

few young juvenile mandibles (MWS 1–10) are present, and a peak of mortality is seen at MWS = 34, a point when the third molar is coming into wear. This corresponds to pigs of approximately three years of age. For about 10% of the pigs, MWS = 41 or greater. These are mature adult pigs, probably including some breeding sows. However, the majority of the Wicken Bonhunt pigs appear to have been male. The upper and lower canine teeth indicate the presence of 817 males and only 510 females.

The maturity of the Wicken Bonhunt pigs can be seen more clearly when the mandibular kill-patterns are compared to the patterns seen at the Early Saxon village of West Stow (n = 315) (Figure 4.2). While both sites show a modal age of death at approximately 3 years (MWS = 31–35), The West Stow assemblage includes a much higher proportion of neonates and young juveniles (MWS = 1–10). The Wicken Bonhunt assemblage includes a higher number of mature and elderly pigs (MWS = 36+). The mandibular evidence, by itself, might be seen as evidence that Wicken Bonhunt served as a site for pig breeding and that many of the younger animals were exported to other sites. The presence of ample pannage for large numbers of pigs within a 3–8km radius of the site would certainly support this interpretation. At the time of the Norman Conquest in the 11th century, Wica and Banhanta were recorded as two separate manors. The Wica manor included woodland with pannage for 100 swine (Morant 1758, 587–8).

The ageing data based on epiphyseal fusion (Table 4.1) present a somewhat different picture. In this table, only the unfused epiphyses of long bone shafts were counted; isolated unfused epiphyses were excluded from consideration. Ages of fusion were based on Silver (1969). The table indicates that between 22% and 38% of the pigs were killed during the first year of life, a figure that is considerably higher than the number of juvenile mortalities indicated by the data on dental eruption and wear. Moreover, the epiphyseal data suggest that a majority of the pigs were killed by 2–2.5 years and that only a small proportion survived for more than 3.5 years.

Age Profiles for Pigs from Wicken Bonhunt and West Stow

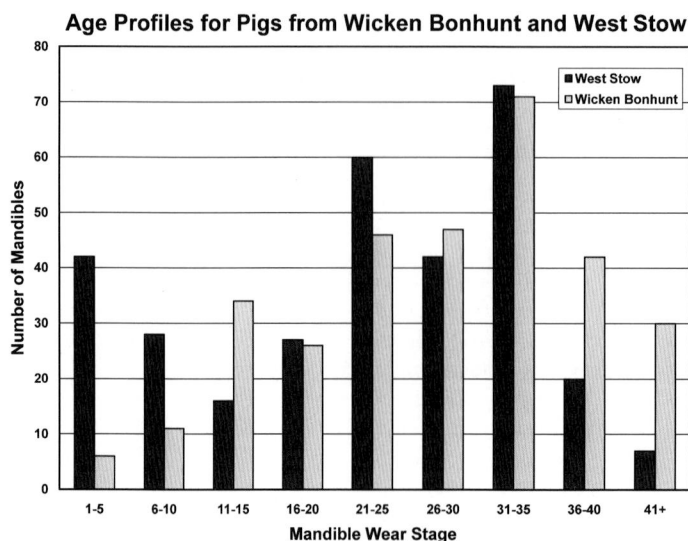

Figure 4.2 Comparison of the kill-patterns for pigs from Wicken Bonhunt and West Stow

26

	N Fused	N Unfused	%Fused	Age of Fusion
Scapula distal	138	72	66	1 yr
Humerus distal	112	69	62	1 yr
Radius distal	85	24	78	1 yr
Total early fusing	**335**	**165**	**67**	
Metacarpus distal	138	190	42	2 yrs
Tibia distal	82	146	36	2 yrs
Metatarsus distal	86	188	31	2 yrs
Calcaneus tuber	4	40	9	2–2.5 yrs
Fibula distal	7	25	22	2.5 yrs
Total middle fusing	**317**	**889**	**26**	
Humerus proximal	12	50	19	3.5 yrs
Radius distal	6	30	17	3.5 yrs
Ulna proximal	4	84	5	3–3.5 yrs
Femur proximal	4	28	13	3.5 yrs
Femur distal	7	37	16	3.5 yrs
Tibia proximal	9	41	18	3.5 yrs
Fibula proximal	0	12	0	3.5 yrs
Total late fusing	**42**	**282**	**13**	

Table 4.1 Epiphyseal fusion data from Middle Saxon pigs from Wicken Bonhunt

This is exactly the *opposite* of the pattern that would be expected if the unfused limb bones of juvenile pigs were preferentially destroyed by taphonomic processes such as carnivore activity (see Brain 1967).

The disparity in the ageing evidence needs explanation. The inhabitants of Wicken Bonhunt may have consumed younger animals, but the meat of many older animals was exported, although the skulls and jaws of these animals remained at the Wicken Bonhunt site. This would also explain the disparity in numbers between the mandibles and the post-cranial skeletal elements. The alternative explanation, that the skulls and mandibles of mature pigs were imported into Wicken Bonhunt, seems less likely since plentiful pannage was available in the area around the site. Even if the skulls and mandibles were imported into Wicken Bonhunt, the data might suggest that the inhabitants were engaged in specialised meat processing, such as the production of head cheese.

The faunal data from Wicken Bonhunt show several important similarities to the data from the Iron Age site of Mount Batten in Dover, a site that Maltby (2006) has argued may have been involved in the long distance trade in pork products. Both sites yielded a high proportion of pig bones, a high number of mandibles and other cranial elements, and a concentration on mandibles from older individuals (Maltby 2006, 119). Maltby has suggested that salt may have been used to preserve and cure the Mount Batten pigs, and a similar argument could be made for the Wicken Bonhunt pigs as well.

The butchery data from Wicken Bonhunt also indicate that specialised pork production was taking place at the site. The mandibles were removed from the skulls in a consistent way. Stevens (1994) reports that 281 of the mandibles show clean chops through the ascending ramus. These chops were always made at a slight angle from the buccal side. In addition, 151 mandibles were split in an anterio-posterior (cranial-caudal) direction. Many of the skulls appear to have been split sagittally for the removal of the brain. The systematic butchery of these cranial elements is consistent with large-scale pork processing.

The ageing and butchery data for pigs from Wicken Bonhunt indicate that the site was not a closed, self-sufficient community. It appears to be a production site that formed part of a broader network of trade and exchange in animal products. The large numbers of skull fragments, mandibles, and loose teeth suggest that large-scale pork production was taking place at Wicken Bonhunt during the Middle Saxon period. Since there is little documentary evidence available for the Bonhunt site, it is not clear whether the pork was exported to contemporary proto-urban centres such as Ipswich and London, whether it was traded to other rural and monastic centres, or whether it was used by local elites for purposes such as feasting, warfare and the support of craft workers and other non-farming specialists. The data cast some doubt on Hodges' (1982, 142) contention that Wicken Bonhunt was a food-rent collection centre. If Wicken Bonhunt were such a collection centre, one would expect a higher proportion of the meaty limb bones of pigs and relatively fewer skulls and jaws. The site of Higham Ferrers in Northamptonshire appears to have served as a food-rent collection centre in the Middle Saxon period. Unlike Wicken Bonhunt, the site produced relatively equal numbers of cattle and pigs during the first half of the 8th century, and a preponderance of cattle during the second half (Evans 2007). Both the dental ageing data and the evidence based on epiphyseal fusion indicate that the vast majority of the Higham Ferrers pigs were slaughtered as juveniles (Evans 2007, 149), a pattern that

Mandible Wear Stages for Brandon Pigs

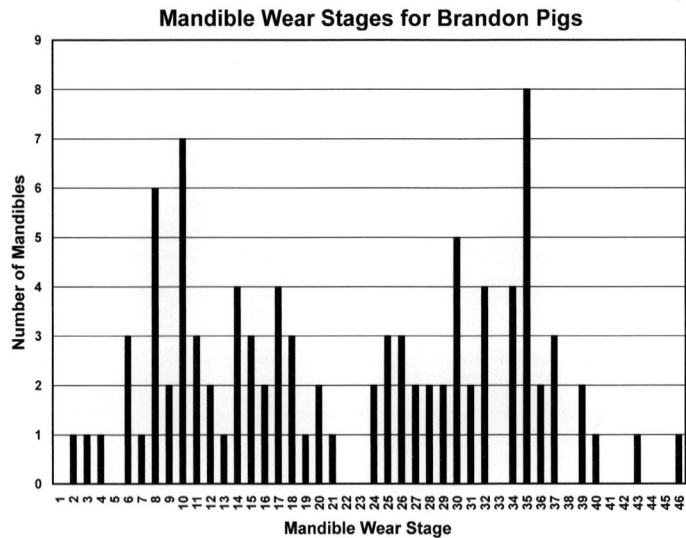

Figure 4.3 Distribution of Mandible Wear Stages for complete pig mandibles from Brandon

is very different from Wicken Bonhunt. In contrast, the ageing and body-part evidence from Wicken Bonhunt suggest that this is a producer rather than a consumer assemblage.

Dobney *et al.* (2007, 146 and fig. 7.33) note that the mandible age wear evidence from Wicken Bonhunt 'shows an extremely similar pattern of slaughter to that from Flixborough'. They further argue that both assemblages show seasonal slaughter of pigs, based on the non-continuous distribution of mandible wear stages (see Figure 4.1). However, the patterns of pig exploitation at the two sites were clearly somewhat different. Dobney *et al.* (2007, 138) note that the epiphyseal fusion data from Phase 2–3a at Flixborough (late 7th to mid-8th century) largely mirror the data based on dental eruption and wear, with a later cull based on older animals. This is clearly not the case at Wicken Bonhunt. The inhabitants of Middle Saxon Flixborough appear to be engaged in pork production for local consumption, while the denizens of

Wicken Bonhunt were producing pork on a much larger scale for export to other sites.

At the Middle Saxon rural site of Brandon, pork production played a relatively minor role in the animal economy. This is not surprising since the Breckland region of Suffolk was never heavily wooded and is far more suited to sheep and cattle husbandry. Humans began to clear the light woodland from the Breckland region in the early Neolithic. The Breckland region had almost no woodland from the Iron Age until the 18th century (Rothera 1998). Not surprisingly, pigs are also third in importance, after cattle and sheep, at the Early Saxon site of West Stow (Crabtree 1990a), and they also play a minor role in the faunal assemblage recovered from the Early Saxon pits and SFBs at Spong Hill in Norfolk (Bond 1995, table 24).

The mandible wear stage data for the pigs from Brandon is summarised in Figure 4.3 which includes only the complete mandibles. While these data show modes of

Age Profiles for Pigs from Brandon

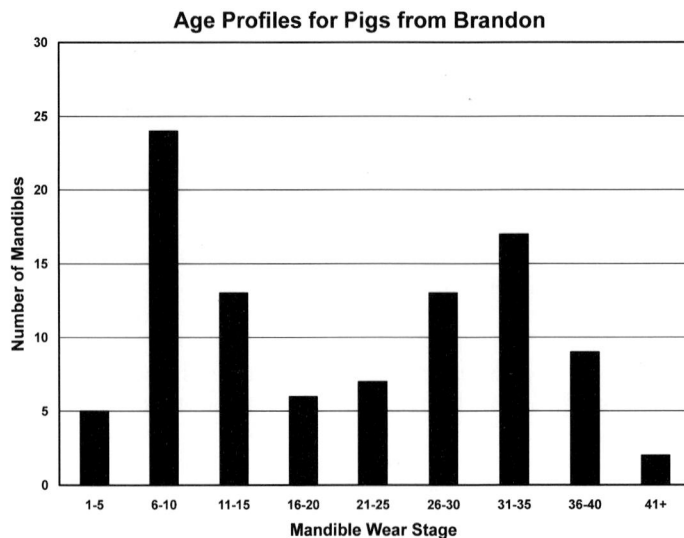

Figure 4.4 Distribution of Mandible Wear Stages in increments of five for pig mandibles from Middle Saxon Brandon

28

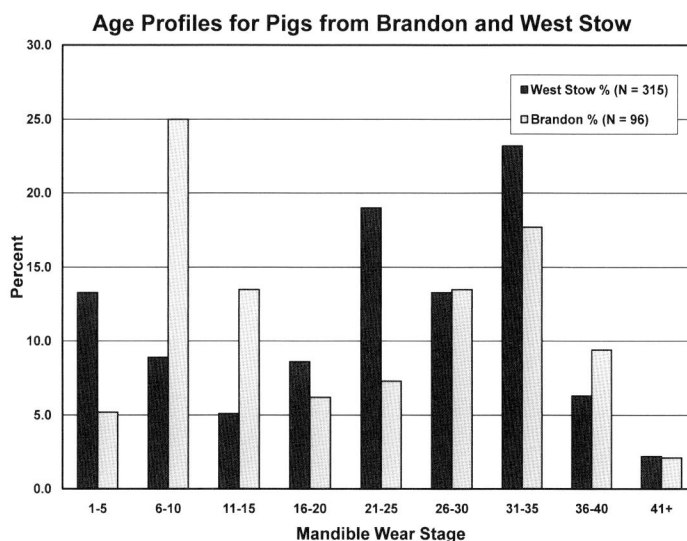

Figure 4.5 Comparison of kill-patterns for pigs from Brandon and West Stow

mortality around MWS = 10 and MWS = 35, pigs of all ages were culled at Brandon.

A larger number of partial and nearly complete mandibles could be assigned an approximate MWS. The data were grouped into classes of five mandible wear stages (Fig. 4.4) to make the information consistent with the ageing data from Early Saxon West Stow. The Brandon data show a mode of mortality at MWS 6–10. These are young, not suckling, pigs with some wear on their first permanent molars. These probably represent seasonal kills of animals that could not be overwintered. A second mode of mortality is seen at MWS 31–35. At this stage the third molar is coming into wear, and the pig is reaching bodily maturity. Continuing to feed a pig after this point will not lead to increased meat yield, so it is reasonable to assume that a high proportion of the pigs will be culled at this stage. Only a small portion of the Brandon pig population survived to advanced ages (MWS = 41+).

When the Brandon mortality profile is compared to the age profile for Early Saxon West Stow (Fig. 4.5), some differences are readily apparent. The West Stow assemblage includes a higher proportion of neonates (MWS = 1–5), while the Brandon assemblage includes a higher percentage of young juveniles (MWS between 6 and 15). Both assemblages include very few old pigs (MWS = 41+). A Kolmogorov-Smirnov test indicates that the differences between the two mortality profiles are significant at the p = 0.05 level but not at the p = 0.025 level.

Ageing data for the Brandon pigs based on epiphyseal fusion are presented in Table 4.2. These data are generally consistent with the ageing evidence based on dental eruption and wear. Based on epiphyseal fusion, just over one-quarter of the pigs were slaughtered in the first year of life. Over 60% of the pigs were slaughtered before they reached 2.5 years of age, and only about 6% survived to more than 3.5 years of age. The ageing data based on both epiphyseal fusion and dental eruption and wear suggest that the Middle Saxons at Brandon kept small numbers of

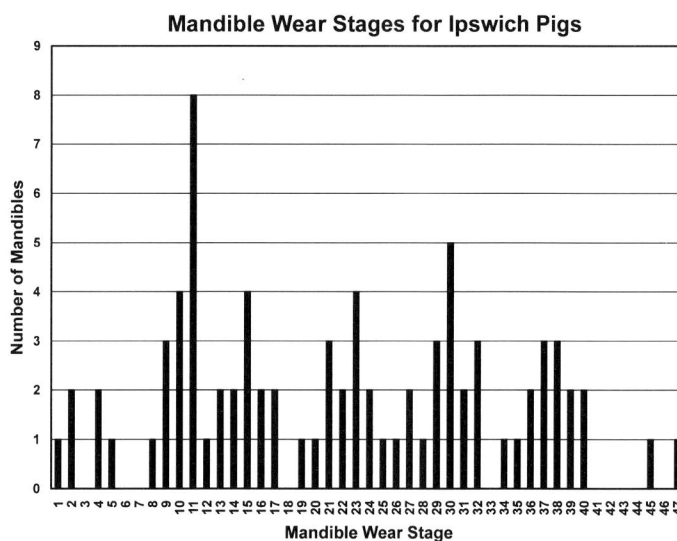

Figure 4.6 Distribution of Mandible Wear Stages for pig mandibles from Middle Saxon Ipswich

29

	Fused	Unfused	Total	% Fused	Age of Fusion
Humerus distal	203	109	**312**	65.1	1 yr
Radius proximal	224	42	**266**	84.1	1 yr
Phalanx 2 Proximal	51	21	**72**	70.8	1 yr
Total early fusing	**478**	**172**	**650**	**73.5**	
Phalanx 1 proximal	69	65	**134**	51.5	2 yrs
Metacarus distal	32	83	**115**	27.8	2 yrs
Tibia distal	164	187	**351**	46.7	2 yrs
Metatarsus distal	25	106	**131**	19.1	2.5 yrs
Metapodium distal	30	60	**90**	33.3	2–2.5 yrs
Calcaneus tuber	11	173	**184**	6.0	2–2.5 yrs
Fibula distal	4	18	**22**	18.2	2.5 yrs
Total middle fusing	**335**	**692**	**1027**	**32.6**	
Ulna proximal	4	103	**107**	3.7	3–3.5 yrs
Ulna distal	3	31	**34**	8.8	3–3.5 yrs
Humerus proximal	5	62	**67**	7.5	3.5 yrs
Radius distal	4	93	**97**	4.1	3.5 yrs
Femur proximal	6	90	**96**	6.3	3.5 yrs
Femur distal	15	126	**141**	10.6	3.5 yrs
Tibia proximal	1	99	**100**	1.0	3.5 yrs
Fibula proximal	0	11	**11**	0.0	3.5 yrs
Total late fusing	**38**	**615**	**653**	**6.3**	

Table 4.2 Epiphyseal fusion data from Middle Saxon pigs from Brandon

	N Fused	N Unfused	%Fused	Age of Fusion
Scapula distal	122	47	72	1 yr
Humerus distal	120	0	100	1 yr
Radius distal	119	21	85	1 yr
Total early fusing	**361**	**68**	**84**	
Metacarpus distal	40	79	34	2 yrs
Tibia distal	55	66	45	2 yrs
Metatarsus distal	28	88	24	2 yrs
Calcaneus tuber	8	83	9	2–2.5 yrs
Fibula distal	0	9	0	2.5 yrs
Total middle fusing	**131**	**325**	**29**	
Humerus proximal	1	8	11	3.5 yrs
Radius distal	17	63	21	3.5 yrs
Ulna proximal	16	90	15	3–3.5 yrs
Femur proximal	0	9	0	3.5 yrs
Femur distal	2	51	4	3.5 yrs
Tibia proximal	3	23	12	3.5 yrs
Fibula proximal	0	2	0	3.5 yrs
Total late fusing	**39**	**246**	**14**	

Table 4.3 Epiphyseal fusion data from Middle Saxon pigs from Ipswich

pigs for home consumption. Excess young pigs were not overwintered; they were slaughtered during their first year of life. Most of the rest of the pigs were slaughtered by the time they reached bodily maturity. Only a small number of elderly animals survived; they probably represent a small breeding stock.

The ageing data for pigs from Ipswich can be used to determine whether the Middle Saxon inhabitants of Ipswich were provisioned with animals of selected age classes, or whether the residents of Ipswich may have kept a small number of pigs for home consumption. In early urban centres in the Ancient Near East, city dwellers were often provisioned with animals of selected age classes. For example, Zeder (1988; 1991) has shown that the late 4th- and 3rd-millennium BC inhabitants of Tal-e Malyan were provisioned with caprines of selected age classes. On the other hand, Milne and Crabtree (2001) demonstrated that the 19th-century inhabitants of the Five Points neighbourhood in New York City raised pigs for personal consumption. The Five Points faunal assemblages associated with immigrant Irish-Americans included substantial numbers of neonatal and young juvenile pigs.

The age profile for the pigs from Middle Saxon Ipswich based on dental eruption and wear is shown in Figure 4.6. The faunal sample includes a small number of suckling pigs (MWS = 1–5) which were likely reared in and around Ipswich. A substantial number of pigs were killed around MWS 11. This mode of mortality probably represents the seasonal slaughter of juvenile pigs that were not overwintered. The mortality profile suggests seasonal culls in the first, second and third years of life, with a small population surviving to more than old age (MWS = 41+). This pattern of mortality is also seen in the epiphyseal data (Table 4.3). Taken together, the data based on dental eruption and wear and those based on epiphyseal fusion of the limb bones suggest that the inhabitants of Middle Saxon Ipswich were raising some pigs for home consumption.

Cattle

Cattle are multi-purpose animals, providing meat, milk, and traction. Analyses of age profiles can help determine whether Middle Saxon cattle were raised for specific purposes, or whether they were used for a range of different primary and secondary products. Mortality profiles for cattle from rural sites may reveal Middle Saxon husbandry strategies, while those from urban sites, such as Ipswich, may reveal whether the inhabitants of these urban communities were provisioned with cattle of specific age classes.

Detailed studies of traction pathologies (Bartosiewicz *et al.* 1997) can also be used to determine whether cattle were used to draw carts and ploughs. Unfortunately, the methods used for identifying and recording these pathologies were developed after the animal bones from Brandon, Ipswich and Wicken Bonhunt were analysed in the early 1990s. Therefore, a quantitative study of traction pathologies cannot be carried out, but some more qualitative observations on traction pathologies can be made.

At Wicken Bonhunt, cattle are the second most common species after pigs. The age profiles for cattle based on dental eruption and wear, following Grant (1982), are shown in Figure 4.7. The data clearly show that the majority of the cattle recovered from Wicken Bonhunt were mature-to-elderly animals. Almost no neonatal and very few juvenile animals were recovered from the site. The paucity of very young animals, a pattern that differs from the age profiles seen for cattle at Early Saxon West Stow (Crabtree 1990a, 69–75) and Middle Saxon Flixborough (Dobney *et al.* 2007, 129, fig. 7.21), suggests that cattle *rearing* may not have played a major role in the Middle Saxon economy of Wicken Bonhunt. It is possible that some of the Wicken Bonhunt cattle were raised elsewhere and brought to the site as adults for meat or for use as traction animals. If cattle were being reared at Wicken Bonhunt, then the excess young animals must have been sent elsewhere. The focus of cattle breeding then would be on older animals that may have been used for traction as part of more intensive agricultural production.

The predominance of mature and elderly cattle at Wicken Bonhunt can be seen more clearly when the mandibles are grouped into broader age classes (Figure 4.8), following Bourdillon and Coy (1980, 87). As can be seen in Figure 4.8, 75 of the 99 ageable mandibles fall into

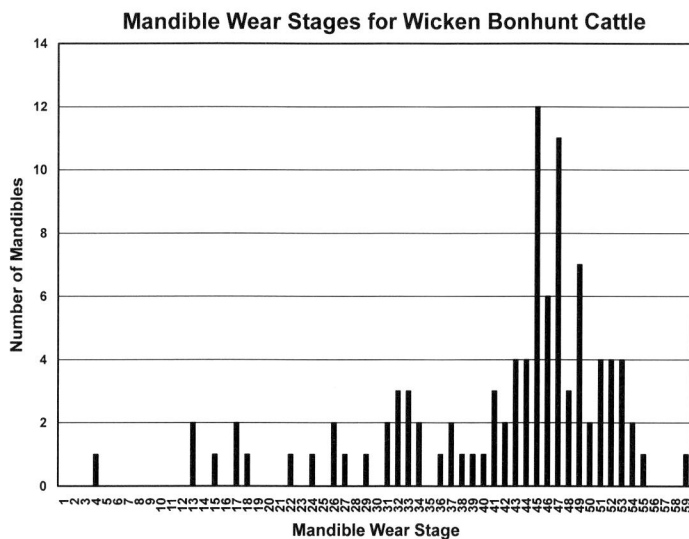

Figure 4.7 Distribution of Mandible Wear Stages for cattle mandibles from Middle Saxon Wicken Bonhunt

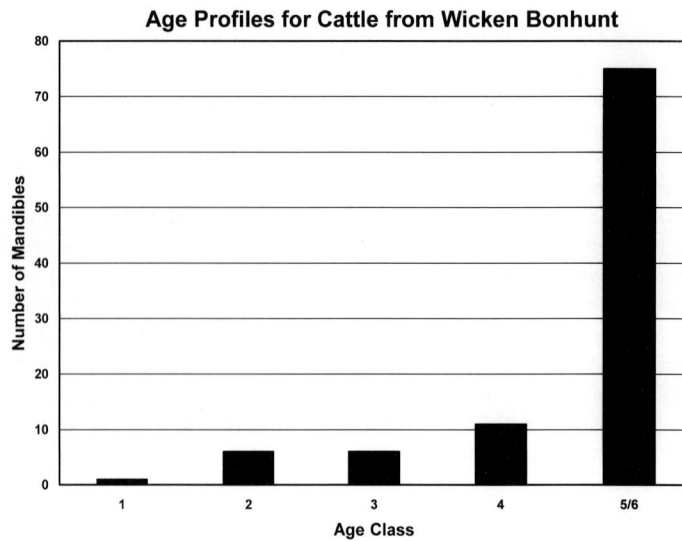

Age Profiles for Cattle from Wicken Bonhunt

Figure 4.8 Distribution of age classes for cattle from Wicken Bonhunt

the oldest age classes. If a mandible wear stage of about 40 is taken to be the equivalent of approximately 5 years of age, then 72 (72%) of the cattle were at least 5 years of age when they were slaughtered.

The ageing data for the Wicken Bonhunt cattle based on epiphyseal fusion are shown in Table 4.4. These data are generally consistent with the data based on dental eruption and wear. The epiphyseal data indicate that less than 10% of the cattle were killed during the first 18 months of life and that fewer than 15% were killed by 2–3 years of age. With the exception of the anomalous data for the proximal ulna, the epiphyseal evidence indicates that the majority of the cattle (56–83%) survived for more than 4 years.

The near absence of young animals suggests that dairying probably did not play a major role in the Middle Saxon economy of Wicken Bonhunt. McCormick (1992)

has argued that primitive cattle required the presence of their calves to let down their milk and young calves would have been weaned early, rather than slaughtered, in the early medieval period. However, Mulville *et al.* (2005) have used the exceptionally well preserved faunal data from the Scottish Isles to show that the presence of substantial numbers of neonates can be seen as a sign of a dairying economy. The mature cattle were certainly used for meat, and some of these animals were probably also used for traction. The metrical data (Chapter 5) indicate that many of these adult animals were oxen rather than cows. In addition, there is some pathological evidence to support the use of cattle for traction purposes. Five first phalanges show some evidence for traction pathology including lipping and distortion of the proximal joint surface and exostosis and expansion of the distal joint surface. These data further support the idea that the

	N. Fused	*N. Unfused*	*% Fused*	*Age of Fusion*
Humerus distal	117	12	91	1–1.5 yrs
Radius proximal	75	3	96	1–1.5 yrs
Scapula distal	89	2	98	1–1.5 yrs
Total Early Fusing	**281**	**17**	**94**	
Tibia distal	138	25	85	2–2.5 yrs
Metacarpus distal	49	5	91	2–2.5 yrs
Metatarsus distal	57	9	86	2.5–3 yrs
Total Middle Fusing	**244**	**39**	**86**	
Calcaneus tuber	59	28	68	3–3.5 yrs
Femur proximal	35	14	71	3.5 yrs
Humerus proximal	30	6	83	3.5–4 yrs
Radius distal	74	23	76	3.5–4 yrs
Ulna proximal	3	14	18	3.5–4 yrs
Femur distal	25	21	57	3.5–4 yr
Tibia proximal	28	17	62	3.5–4 yrs
Total Late Fusing	**254**	**123**	**67**	

Table 4.4 Epiphyseal fusion data from Middle Saxon cattle from Wicken Bonhunt

Mandible Wear Stages for Brandon Cattle

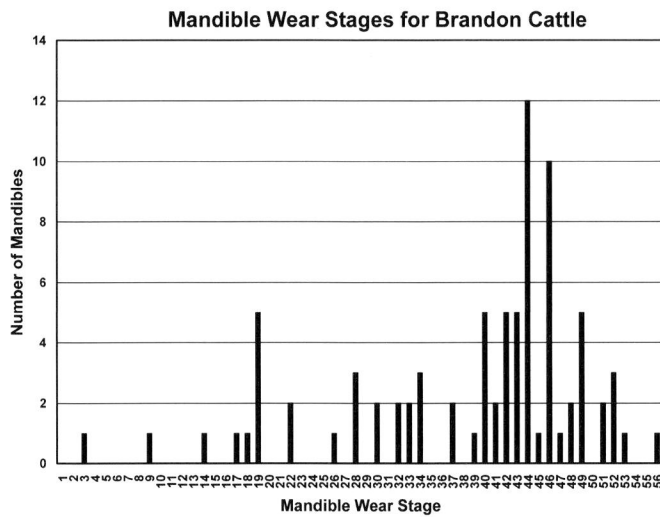

Figure 4.9 Distribution of Mandible Wear Stages for cattle mandibles from Middle Saxon Brandon

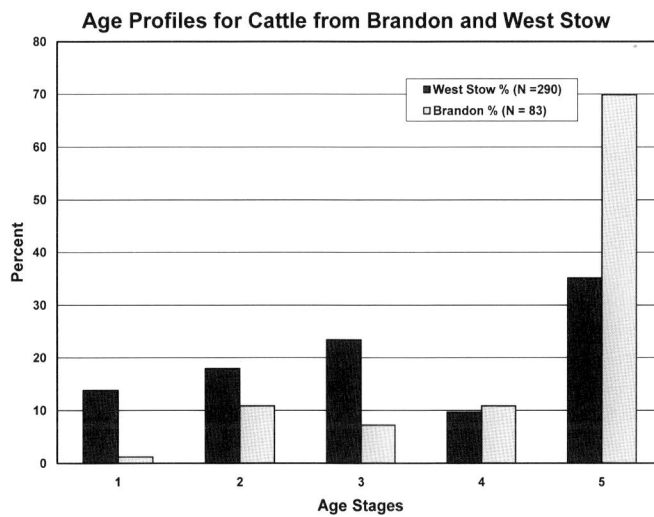

Age Profiles for Cattle from Brandon and West Stow

Figure 4.10 Comparison of age profiles for Middle Saxon cattle from Brandon with Early Saxon cattle from West Stow

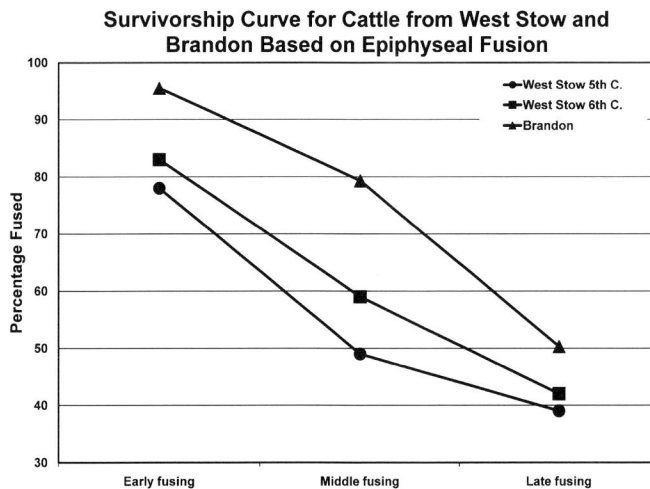

Survivorship Curve for Cattle from West Stow and Brandon Based on Epiphyseal Fusion

Figure 4.11 Age profiles based on epiphyseal fusion for cattle from Middle Saxon Brandon and 5th- and 6th-century contexts at West Stow

Mandible Wear Stages for Ipswich Cattle

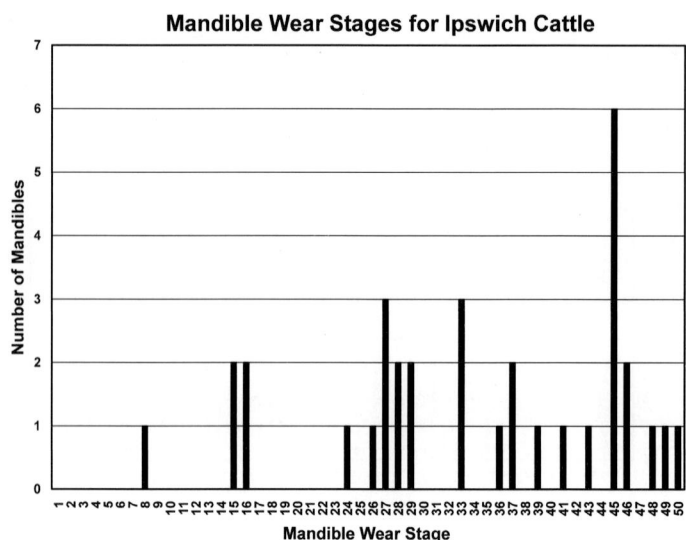

Figure 4.12 Distribution of Mandible Wear Stages for cattle mandibles from Middle Saxon Ipswich

Wicken Bonhunt economy was geared toward the production of pork and that cattle were used primarily for traction and transport.

The ageing data for cattle at Brandon based on dental eruption and wear present a broadly similar picture (Figure 4.9). The distribution indicates that the majority of cattle from Brandon survived to maturity and that very few cattle were killed during the first year of life. When these mandible wear stages are combined into broader age classes and compared to the cattle kill-patterns from West Stow (Figure 4.10), it is apparent that the Brandon assemblage includes a higher proportion of mature animals. A Kolmogorov-Smirnov test was used to compare the Brandon and the West Stow cattle age distributions. The differences between the two assemblages are significant at the $p = 0.001$ level.

The kill-patterns seen for the dental ageing from Brandon are paralleled in epiphyseal ageing evidence (Table 4.5). These data show that only about 5% of the cattle were killed during the first 18 months of life. 80% of

the Brandon cattle survived to 3 years of age, and just over half the cattle survived to age 4. Figure 4.11 shows the Brandon kill-pattern for epiphyseal fusion compared to the patterns seen at West Stow Phase 1 (5th century) and Phase 2 (6th century). The early fusing elements are those that fuse by 1.5 years; the middle fusing elements are those that fuse by 3 years; and the late fusing elements are those that fuse by 4 years. The figure shows that the Brandon cattle consistently survived to older ages than the West Stow cattle did.

The reasons for the differences in the kill patterns between the Early and Middle Saxon periods are not entirely clear. It is certainly possible that Brandon was provisioned with market-age cattle from other Anglo-Saxon sites, possibly in the form of rent or tribute. It is also possible that the Brandon assemblage includes a larger proportion of draught cattle that were slaughtered after they had been used for traction. However, there is little pathology on the Brandon cattle that can be directly attributed to their use as traction animals.

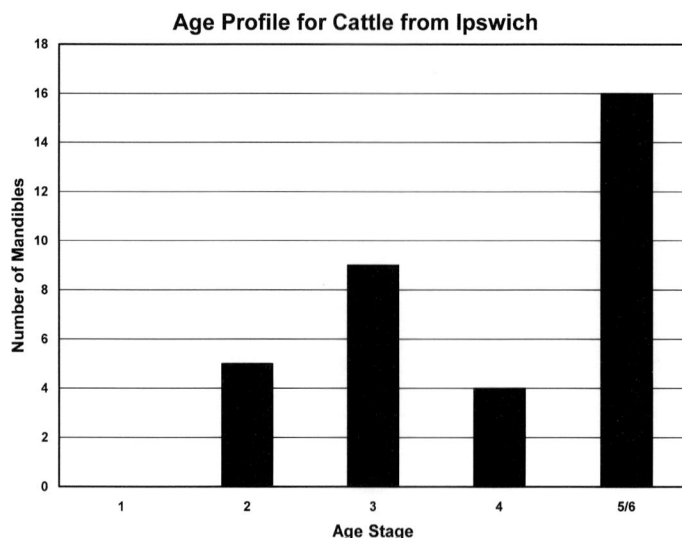

Age Profile for Cattle from Ipswich

Figure 4.13 Distribution of broader age classes for Middle Saxon cattle from Ipswich

34

	N. Fused	N. Unfused	% Fused	Age of Fusion
Humerus distal	283	22	92.8	1–1.5 yrs
Radius proximal	332	7	97.9	1–1.5 yrs
First Phalanx proximal	570	37	94.5	1.5 yrs
Second Phalanx proximal	330	6	98.2	1.5 yrs
Total Early Fusing	**1515**	**72**	**95.5**	
Tibia distal	294	69	81.0	2–2.5 yrs
Metacarpus distal	184	37	83.3	2–2.5 yrs
Metatarsus distal	174	52	77.0	2.5–3 yrs
Metapodium distal	51	25	67.1	2–3 yrs
Total Middle Fusing	**703**	**183**	**79.3**	
Calcaneus tuber	112	95	54.1	3–3.5 yrs
Femur proximal	152	120	55.9	3.5 yrs
Humerus proximal	17	26	39.5	3.5–4 yrs
Radius distal	94	73	56.3	3.5–4 yrs
Ulna proximal	16	27	37.2	3.5–4 yrs
Ulna distal	0	4	0.0	3.5–4 yrs
Femur distal	53	93	36.3	3.5–4 yrs
Tibia proximal	62	61	50.4	3.5–4 yrs
Total Late Fusing	**506**	**499**	**50.3**	

Table 4.5 Epiphyseal fusion data from Middle Saxon cattle from Brandon

Not everyone at Brandon may have had equal access to beef. A metre-by-metre plot of the distribution of the cattle remains shows a concentration between buildings 8893 and 8927. These buildings have been assigned to Middle Saxon Phase 2.3 and are the final buildings in that region of the site. These buildings appear to have been occupied by higher status individuals, and the dietary differences may reflect differences in status and wealth at the site.

The symbolic and economic roles of cattle may also have changed between the Early and Middle Saxon periods in East Anglia. Cattle were symbols of wealth in Iron Age and Migration Period sites in the British Isles and on the European continent (see, for example, Crabtree 1986 and references therein), and they may have continued to serve as prestige items at sites such as West Stow. In terms of NISP, cattle are less important at Brandon than they are at Early Saxon West Stow. It is possible that by the Middle Saxon period, cattle were no longer primarily valued as symbols of wealth. They became just another form of food and capital, used to pull carts and ploughs and slaughtered when they were no

	N. Fused	N. Unfused	% Fused	Age of Fusion
Humerus distal	165	13	93	1–1.5 yrs
Radius proximal	112	2	98	1–1.5 yrs
Scapula distal	93	3	97	1–1.5 yrs
Total Early Fusing	**370**	**18**	**95**	
Tibia distal	120	44	73	2–2.5 yrs
Metacarpus distal	82	26	76	2–2.5 yrs
Metatarsus distal	79	28	84	2.5–3 yrs
Total Middle Fusing	**281**	**98**	**74**	
Calcaneus tuber	73	56	57	3–3.5 yrs
Femur proximal	5	5	50	3.5 yrs
Humerus proximal	1	2	33	3.5–4 yrs
Radius distal	49	37	57	3.5–4 yrs
Ulna proximal	6	23	21	3.5–4 yrs
Femur distal	15	9	62	3.5–4 yrs
Total Late Fusing	**149**	**132**	**53**	

Table 4.6 Epiphyseal fusion data from Middle Saxon cattle from Ipswich

longer needed for work or other purposes. McCormick (2008) has made a similar argument for cattle in early medieval Ireland.

Although Ipswich was a proto-urban site during the Middle Saxon period, the ageing patterns seen for cattle mirror patterns seen at the rural sites of Brandon and Wicken Bonhunt. The distribution of mandible wear stages for the Middle Saxon cattle from Ipswich is shown in Figure 4.12, and the data are grouped into broader age classes following Bourdillon and Coy (1977, 25) in Figure 4.13. The age distributions show that many, but not all, the cattle survived to advanced years. While the modal mortality stage is seen at MWS = 45, the collection also includes a number of mandibles from juvenile and adolescent animals.

The data on epiphyseal fusion (Table 4.6) indicate that only about 5% of the cattle were slaughtered by 18 months of age, but about one-quarter were killed by the 3 years. The epiphyseal data suggest that about half the cattle survived to more than 4 years of age. If we assume that a mandible wear stage of 40 corresponds to about 5 years of age, then it appears that about half the cattle survived for more than five years. Metrical data (see Chapter 5) indicate that some of these adult cattle may have been oxen. The paucity of very young cattle bones suggests that few, if any, cattle were actually reared at Ipswich. It is far more likely that the denizens of Ipswich were provisioned with cattle from the surrounding countryside. These include both market-age cattle and older animals which were no longer useful for traction purposes.

Sheep

Sheep play a relatively minor role in the rural economy at Wicken Bonhunt. They are third in importance, following pigs and cattle, on the basis of NISP. Ageing data can help to reveal the role that sheep played in Wicken Bonhunt's economy. The distribution of the mandible wear stages for the Wicken Bonhunt sheep are shown in Figure 4.14; in Figure 4.15 they are grouped into broader age classes, following Bourdillon and Coy (1977; 1980) and Crabtree

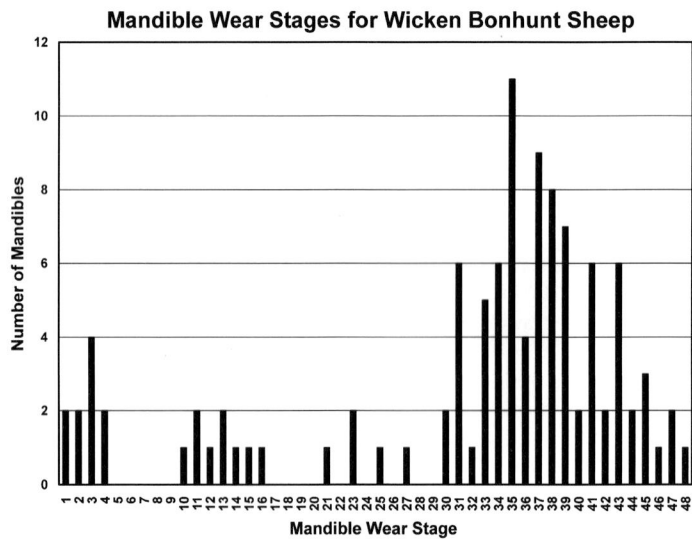

Figure 4.14 Distribution of Mandible Wear Stages for sheep mandibles from Middle Saxon Wicken Bonhunt

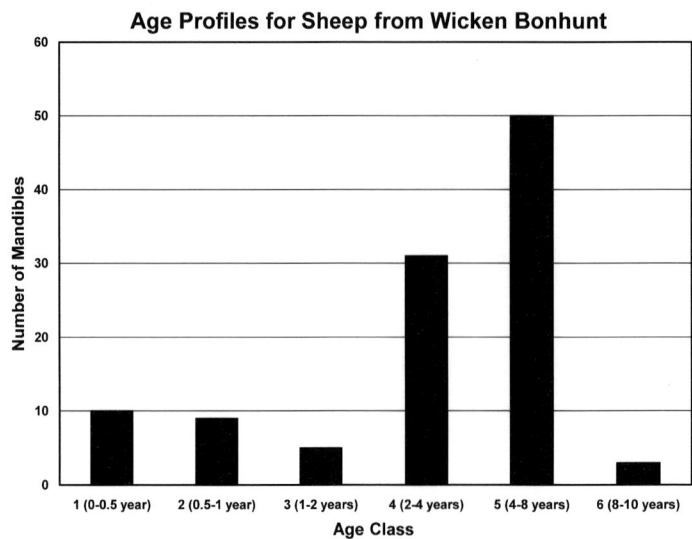

Figure 4.15 Distribution of broader age classes for Wicken Bonhunt sheep

	N. Fused	N. Unfused	% Fused	Age of Fusion
Scapula distal	117	26	82	6–8 months
Humerus distal	120	5	96	10 months
Radius proximal	153	4	96	10 months
First Phalanx proximal	29	1	97	13–16 months
Total Early Fusing	**419**	**36**	**92**	
Tibia distal	212	27	89	1.5–2 yrs
Metacarpus distal	10	3	77	1.5–2 yrs
Metatarsus distal	23	14	62	20–28 months
Total Middle Fusing	**237**	**44**	**84**	
Ulna proximal	11	20	35	2.5 yrs
Femur proximal	12	10	55	2.5–3 yrs
Calcaneus tuber	13	18	42	2.5–3 yrs
Radius distal	53	32	62	3 yrs
Humerus proximal	12	20	38	3–3.5 yrs
Femur distal	10	2	77	3–3.5 yrs
Tibia proximal	26	45	37	3–3.5 yrs
Total Late Fusing	**137**	**147**	**48**	

Table 4.7 Epiphyseal fusion data from Middle Saxon sheep from Wicken Bonhunt

(1990a, table 46). Although the sample of ageable mandibles is relatively small (N = 108), the Wicken Bonhunt data clearly indicate that relatively few sheep were killed before the age of two years (age classes 1, 2, and 3). The assemblage includes a substantial portion (28.7% of the mandibles) of young adult animals that were killed between 2 and 4 years of age (age class 4). These market-age animals may have provided one or more fleeces before they were slaughtered for meat. The majority of sheep were killed between 4 and 8 years of age (Class 5), but a very few animals survived to advanced years (Class 6). The epiphyseal fusion data (Table 4.7)

also show that few of the Wicken Bonhunt sheep were killed during the first two years of life. The high proportion of adult sheep suggests that wool production may have played a role in the Wicken Bonhunt economy. The role of wool, however, was clearly secondary to pork, since pigs are far more common than sheep at Wicken Bonhunt. Sheep may also have been valued for their manure. Sheep can graze on stubble once the crops have been harvested and provide fertiliser for subsequent crops.

Sheep are the most common animals at Brandon; species ratios based on NISP (Chapter 3) show that sheep substantially outnumber both cattle and pigs at this site.

	N. Fused	N. Unfused	% Fused	Age of Fusion
Humerus distal	1148	105	92	10 months
Radius proximal	746	68	92	10 months
First Phalanx proximal	250	30	89	13–16 months
Total Early Fusing	**2146**	**213**	**91**	
Tibia distal	1152	188	86	1.5–2 yrs
Metacarpus distal	222	52	81	1.5–2 yrs
Metatarsus distal	209	64	77	20–28 months
Total Middle Fusing	**1583**	**304**	**84**	
Ulna proximal	75	85	47	
Femur proximal	185	195	49	2.5–3 yrs
Calcaneus tuber	235	123	66	2.5–3 yrs
Radius distal	222	208	52	3 yrs
Humerus proximal	92	149	38	3–3.5 yrs
Femur distal	168	203	45	3–3.5 yrs
Tibia proximal	120	127	49	3–3.5 yrs
Total Late Fusing	**1097**	**1090**	**50**	

Table 4.8 Epiphyseal fusion data from Middle Saxon sheep from Brandon

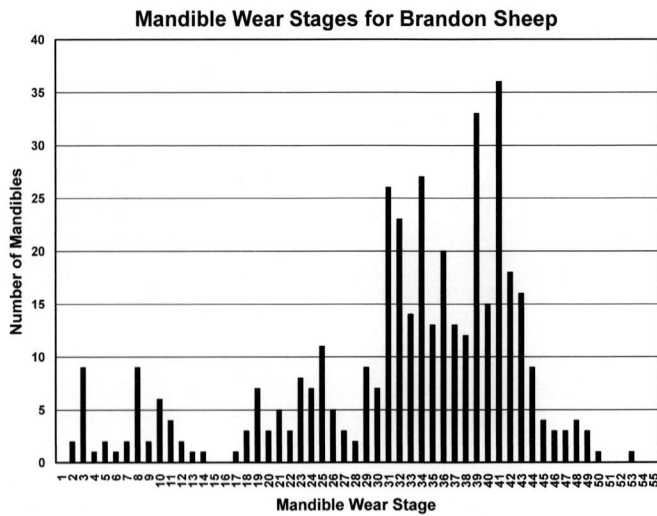

Figure 4.16 Distribution of Mandible Wear Stages for sheep mandibles from Middle Saxon Brandon

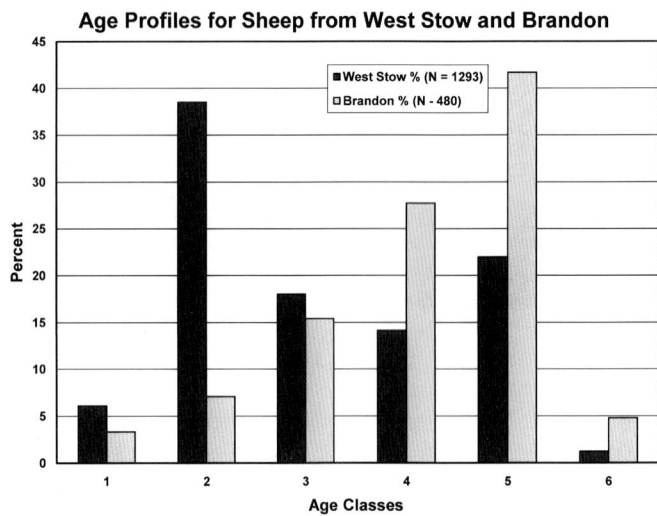

Figure 4.17 Distribution of age classes for sheep, following Bourdillon and Coy (1980), for sheep from Brandon and the original West Stow excavations

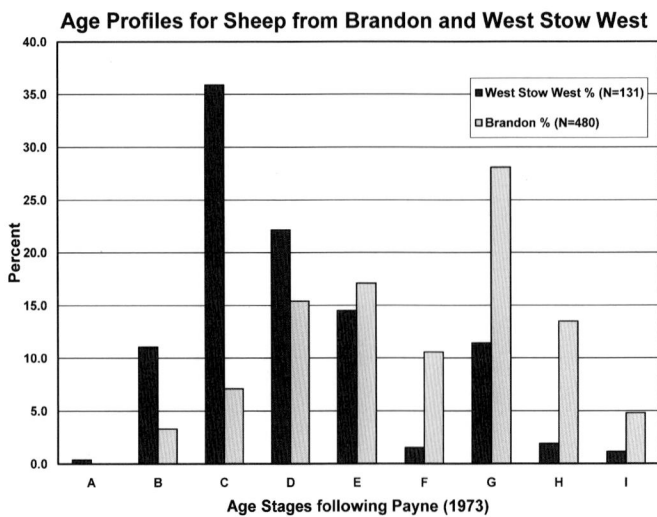

Figure 4.18 Distribution of age classes, following Payne (1973), for sheep from Middle Saxon Brandon and the new excavations at West Stow Visitor Centre

38

The distribution of ageable mandibles from the Middle Saxon contexts at Brandon is shown in Figure 4.16. The age distribution shows that the majority of sheep were culled as adults, including a number of elderly animals with heavily worn molars. A mode of mortality is seen at MWS = 39–41. These were older animals that were probably about 6 years of age when they were slaughtered. The age profiles for the sheep from Brandon and the original excavations at West Stow have been compared in detail elsewhere (Crabtree 2007; Crabtree and Campana n.d.), but Figure 4.17 provides a basic comparison of the culling patterns from Brandon and the original West Stow excavations, following the age classes used by Bourdillon and Coy (1980). In Figure 4.18, the Brandon mandibles and the mandibles from the West Stow Visitor Centre site have been divided into age classes following Payne (1973). These data show that a minority of the Brandon animals were killed in the first two years of life (age classes A–D), a pattern that is very different from the age profile seen at Early Saxon West Stow Visitor Centre. The animals slaughtered between ages 2 and 4 (Stages E and F) would have provided one or more fleeces before they were slaughtered for meat. Many of the Brandon animals survived to advanced ages, while most of the West Stow West animals were slaughtered before they reached maturity. This pattern is also seen in the original West Stow faunal collection (Figure 4.17 and Crabtree 1990a, 86). The mandibular ageing data suggest that the inhabitants of Brandon were focused on wool, rather than meat production. The ageing data also suggest that significant changes took place in sheep husbandry in western Suffolk between the Early and Middle Saxon periods.

The epiphyseal fusion data from Brandon includes identifiable sheep and indeterminate sheep/goat bones. Identifiable goat bones were excluded from the analyses. The data are presented in Table 4.8. Following Silver (1969, 285–6), the epiphyses were divided into early fusing (by 16 months), middle fusing (by 28 months), and late fusing (by 42 months) groups. These data are generally consistent with the mandibular evidence. These data indicate that less than 10% of the Brandon sheep were culled by 16 months, less than 20% were culled by 28 months, and about half survived to more than 3.5 years of age. In contrast, the epiphyseal data from the original West Stow excavations (Crabtree 1990a, table 45) indicate that 15–20% of the West Stow sheep were culled by 16 months; about 40% were culled by 28 months; and only about 30% survived for more than 3.5 years.

As I have argued elsewhere (Crabtree 1995; 2007; Crabtree and Campana n.d.), multiple lines of evidence support the interpretation that the Brandon economy was focused on raising sheep for wool production. The age profiles based on both epiphyseal fusion and dental eruption and wear indicate that the Anglo-Saxon farmers at Brandon focused on older sheep. While ewes are the productive members of dairy herds, wethers (castrated male sheep) are excellent wool producers. Of the 521 sheep horn cores and pelves whose sex could be determined with reasonable certainty, 309, approximately 59%, were males. In addition, remains of flax (Linum usitatissimum), hemp (Cannabis sativa), and dye plants were recovered from the waterfront industrial area at Brandon, suggesting that textile production took place there (Carr et al. 1988). In short, the archaeological, faunal and archaeobotanical data suggest that intensive wool production, as well as textile production, took place at Brandon during the Middle Saxon period.

The ageing data for sheep from the Middle Saxon urban site of Ipswich are somewhat more difficult to interpret. The age profile based on dental eruption and wear is shown in Figure 4.19. The data show a mode of mortality around MWS = 9–12. These probably represent first-year sheep that farmers chose not to overwinter. The distribution includes fewer second-year culls, and a large number of late adolescent to mature animals. When the sheep are divided into broader age classes (Figure 4.20), the pattern of mortality parallels the kill-patterns seen at rural Early Saxon West Stow (original excavations). These data might suggest that the inhabitants of Ipswich were provisioned by relatively unspecialised farmers, similar to those seen at West Stow and other Early Saxon sites in eastern England.

The epiphyseal data (Table 4.9) paint a somewhat different picture. These data suggest that only about 5% of the Ipswich sheep were slaughtered during the first 16 months of life and that only about one-quarter were slaughtered by 28 months. The epiphyseal data suggest

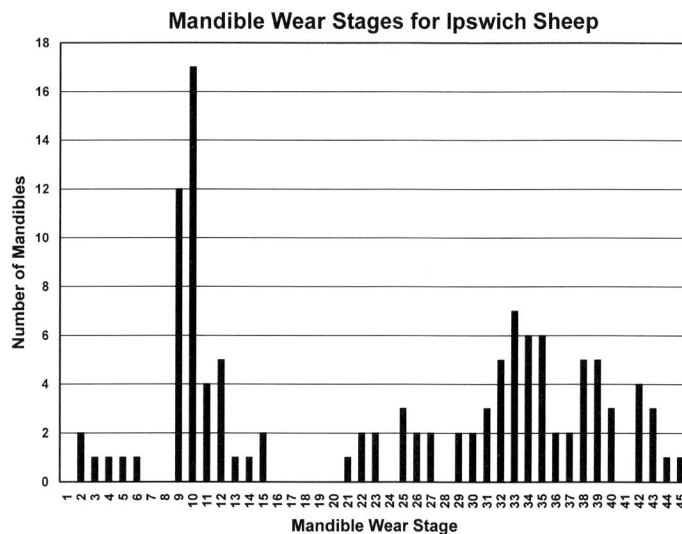

Figure 4.19 Distribution of Mandible Wear Stages for sheep mandibles from Middle Saxon Ipswich

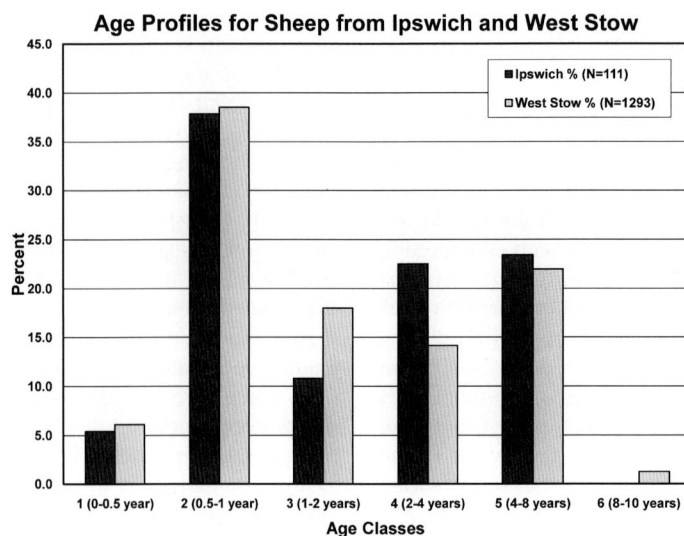

Age Profiles for Sheep from Ipswich and West Stow

■ Ipswich % (N=111)
□ West Stow % (N=1293)

Age Classes: 1 (0-0.5 year), 2 (0.5-1 year), 3 (1-2 years), 4 (2-4 years), 5 (4-8 years), 6 (8-10 years)

Figure 4.20 Distribution of broader age classes for sheep from Middle Saxon Ipswich and from the original excavations at Early Saxon West Stow

that just over half the sheep consumed by the inhabitants of Ipswich came from sheep that were more than 3.5 years old.

How can we reconcile the epiphyseal and the dental evidence? Certainly, one possibility is that the fragile unfused epiphyses of the first-year culls were differentially destroyed by carnivore activity. Another possibility is that mature sheep were slaughtered elsewhere and sold as joints of mutton in markets. Both processes could have been at work here.

The Ipswich faunal assemblage may present some unique contextual issues as well. While the Brandon and Wicken Bonhunt faunal assemblages represent substantial horizontal excavations of large rural sites, the Ipswich assemblage is composed of a number of different small faunal samples derived from different sites within the *emporium* (see Appendix for details). Cranial and

post-cranial remains may have been disposed of in different ways, and inhabitants of different parts of the *emporium* may have been supplied with meat from lambs and sheep of different ages. Unfortunately, the individual samples are so small that it is not possible to carry out a detailed contextual analysis of the faunal remains from each site.

III. Economic Interpretations

The ageing data from West Stow (Crabtree 1990a), Kilham in Yorkshire (Archer 2003), and the smaller rural sites in eastern England suggest that Early Saxon animal husbandry was largely unfocused and non-specialised. Sheep were raised for meat and possibly milk. While there is clear artefactual evidence for textile production, such as loom weights and spindle whorls, from both Kilham and

	N. Fused	N. Unfused	% Fused	Age of Fusion
Scapula distal	93	3	97	6–8 months
Humerus distal	165	13	93	10 months
Radius proximal	112	2	98	10 months
Total Early Fusing	**370**	**18**	**95**	
Tibia distal	120	44	73	1.5–2 yrs
Metacarpus distal	82	26	76	1.5–2 yrs
Metatarsus distal	79	28	74	20–28 months
Total Middle Fusing	**281**	**98**	**74**	
Ulna proximal	6	23	21	2.5 yrs
Femur proximal	5	5	50	2.5–3 yrs
Calcaneus tuber	73	56	57	2.5–3 yrs
Radius distal	49	37	57	3 yrs
Humerus proximal	1	2	33	3–3.5 yrs
Femur distal	15	9	62	3–3.5 yrs
Total Late Fusing	**149**	**132**	**53**	

Table 4.9 Epiphyseal fusion data from Middle Saxon sheep from Ipswich

West Stow (Hunter-Mann 2001, 20; West 1985, 138–9), the zooarchaeological data provide no evidence for specialised wool production. Cattle were raised for both meat and milk, and the palaeopathological evidence suggests that some of the cattle were also used for traction purposes. The data from Kilham, West Stow, and the smaller sites indicate that Early Saxon farmers focused on cattle- and sheep-rearing. Pigs are always third in importance, as measured by NISP. While there is very limited evidence for small-scale trade in items such as marine fish (see, for example, Crabtree 1990a, table 20), the overall impression is one of subsistence farming and relative economic self-sufficiency.

The data from both Brandon and Wicken Bonhunt suggest that fundamental economic changes took place between the Early and Middle Saxon periods in eastern England. The species ratios, body part distributions (Chapter 3), and ageing data suggest that the Wicken Bonhunt settlement was engaged in large-scale pork production for export to other sites. Brandon is located in the Breckland region of western Suffolk, a region that is very suitable for sheep rearing. The species ratios, sex ratios, and ageing, along with the archaeological and archaeobotanical evidence, strongly suggest that the inhabitants of Brandon were engaged in large-scale wool production. Interestingly, both Brandon and Wicken Bonhunt appear to be wealthy, high-status estate centres.

While not all Middle Saxon rural settlements show evidence for increasing specialisation in animal husbandry practices, there is some evidence for contemporary changes at the settlement of Quarrington in Lincolnshire (Taylor 2003). The site has yielded structures and features that can be dated to both the Early and Middle Saxon periods. The faunal assemblage (Rackham 2003) is dominated by the remains of cattle, followed by sheep, and small numbers of pig bones. The proportion of sheep, relative to cattle and pigs, rises from 25.2% to 34.6% between the Early and Middle Saxon periods. Rackham (2003, 271) has argued that the increasing numbers of sheep may be related to changes in farming intensity and management, and he notes the importance of sheep for meat, milk, wool, and fertiliser. Rackham (2003, 271) documents a shift in the slaughter pattern for sheep from a 'non-focused management of a largely subsistence character' during the Early Saxon period to a more structured pattern with fewer young sheep and a concentration of third- to fifth-year culls during the Middle Saxon period. He suggests that this more focused cull pattern may reflect either the commercial exploitation of these sheep for wool or demands that resulted from newly-created lordships. Either way, the Quarrington data suggest that fundamental social and economic changes took place between the Early and Middle Saxon periods.

At the same time, we see the emergence of the first post-Roman urban centres in Britain and north-west Europe. Ipswich was founded in the early 7th century, and developed as a centre for craft production and regional and international trade during the Middle Saxon period. Since Ipswich was home to workers engaged in pottery production and other craft activities, the question of how these non-agricultural workers obtained meat and other animal products is critical to our understanding of the development of this *emporium*. The ageing data suggest that the inhabitants of Ipswich obtained meat in different

ways. The presence of neonatal and young juvenile pigs in the Middle Saxon assemblages from Ipswich suggests that some pigs were raised in and around the *emporium* for local consumption. Cattle and sheep, on the other hand, appear to have been raised outside the town. Ageing data based on both dental eruption and wear and epiphyseal fusion indicate that the inhabitants of Ipswich were provisioned both with market-aged cattle and with cattle that were no longer useful for traction and transport. The data for sheep are a bit more equivocal. The mandibular data indicate that the inhabitants of Ipswich were provisioned both with market age and older sheep and with first-year culls. The epiphyseal data, on the other hand, suggest that the inhabitants of Ipswich were provisioned primarily with older animals. Not surprisingly, there is no evidence to suggest that sheep were reared in Ipswich.

The critical question for medieval archaeologists is whether there is a relationship between the development of the *emporia* and the apparent changes in rural animal husbandry that take place between the Early and Middle Saxon periods. This is part of a larger theoretical issue surrounding the origin of urban societies. Seventy-five years ago Childe (1936, see also Childe 1950) argued that the production of an agricultural surplus, what Childe termed the social surplus, was a critical factor in the development of complex, urban societies. I have previously suggested (Crabtree 1996) that there was a relationship between the rise of the *emporia* and the increasing specialisation seen in Middle Saxon animal husbandry. However, the process may not be so simple, and multiple explanations are possible for the increasingly focused and specialised animal husbandry practices seen at sites such as Brandon and Wicken Bonhunt.

Hodges (1982) is one of the archaeologists who have suggested that there is a direct relationship between the rise of the *emporia*, state formation, and the transformation of the Anglo-Saxon rural economy. Unfortunately, as noted above, his interpretation of Wicken Bonhunt as a food-rent collection centre is not borne out by the zooarchaeological data. Early historical documents indicate that the Anglo-Saxon landscape was organised to support itinerant royal courts (Blair 2005, 252). If Wicken Bonhunt were a food-rent collection centre, we would expect to see evidence for consumption and feasting, rather than specialised production.

There are other problems with the notion that the rise of the *emporia* led to a rearrangement of rural settlement in order to produce agricultural surpluses (Hinton 1990, 58). These are discussed in some detail by Moreland (2000). One of the critical issues is a chronological one. The settlement shift that led to the foundation of sites like Brandon and the abandonment of Early Saxon sites such as West Stow appears to have begun before the end of the 7th century. The establishment of the Ipswich pottery industry, on the other hand, may have taken place as late as AD 700–720 (Blinkhorn 1999), and the major expansion of the site took place during the 8th century.

Blair (2005) has suggested that monasticism played a major role in the reorganisation of the Middle Saxon countryside. He argues that the Middle Saxon 'sites which were the most highly developed, lasted longest, and yield the widest and richest assemblages of finds bear a strongly monastic stamp' (Blair 2005, 211). Monasteries were in a unique position to develop agricultural surpluses because of their control over land and their access to inexpensive labour from low-status, quasi-monastic workers. Blair identifies both Brandon and Flixborough in Lincolnshire as monastic communities and argues that these communities were engaged in the production of agricultural surpluses and the conspicuous consumption of luxuries and feasting. While this explanation may be appropriate for Brandon, it certainly cannot explain the development of intensified pork production at Wicken Bonhunt, a wealthy but secular Middle Saxon settlement. In addition, the monastic model does not fully explain the economic changes seen at Flixborough. Detailed archaeological and zooarchaeological analyses have shown that hunting, feasting, and conspicuous consumption are characteristic of the 8th century at Flixborough, when the site was under secular control. A decline in conspicuous consumption and increasing craft activity are seen during the 9th century, when Flixborough came under monastic control (Loveluck 2007, 148–54).

A third possibility is that the switch to more specialised production represents decisions made by individual local estate centres, and that these changes began to take place before the rise of the *emporia* in the 8th century. These estate centres, both secular and ecclesiastical, may have become foci of specialised production and trade. Moreland (2000, 69) has suggested that production, consumption and trade are intimately linked in early medieval economies and that the traditional focus on trade has led archaeologists to see the *emporia* and the king as having primary roles in the transformation of the economy during the Middle Saxon period. The data from Brandon and Wicken Bonhunt suggest that a more heterarchical model may well explain the rise of specialised production and exchange during the long 8th century. Individual estate centres may have made decisions to engage in specialised production for exchange and may have even been sites for periodic markets and fairs (see Crabtree 2010b).

Chapter 5. Osteometric Analyses

I. Introduction

Measurements taken on archaeologically-recovered animal remains can contribute to our understanding of Anglo-Saxon animal husbandry in several different ways. First, they can inform us about the size of Middle Saxon animals. By comparing measurements taken on Iron Age, Roman and medieval cattle, we can trace changes in animal sizes through time. In addition, some measurements, such as those taken on cattle metacarpi, can be used to distinguish male, female, and castrated animals. This information is critical to our understanding of animal husbandry since females are the productive members of domestic flocks, supplying milk and offspring. However, castrated male cattle, oxen, are often used for traction and transport purposes, since intact males are generally intractable. In addition, castrated male sheep, wethers, are excellent wool producers. Finally, standardised bone measurements are useful to zooarchaeologists because they are not affected by recovery bias. Over the past 50 years, recovery methods used by archaeologists have changed dramatically. Screening and wet-sieving have replaced hand collection. While poor recovery methods will affect body-part distributions and species ratios since small bones are more likely to be missed by hand-collectors, measurements from hand-collected assemblages can be compared directly to measurements from assemblages that were collected using modern recovery methods.

One of the most important features of the Wicken Bonhunt, Ipswich, and Brandon faunal assemblages is that they yielded large numbers of measurable specimens. While smaller collections require analysts to combine different measurements (using a standard animal and a log scale) to trace changes in animal size through time, the faunal samples from the Middle Saxon sites from East Anglia are large enough that individual measurements can be compared directly. In addition, the large numbers of complete long bones permit the calculation of withers heights for several species (following the recommendations of von den Driesch and Boessneck 1974). As noted in the introductory chapter, all the measurements discussed in this chapter were taken following the recommendations of von den Driesch (1976). In this chapter, the measurements taken on Middle Saxon cattle, sheep, pigs, horses, dogs and chickens will be presented.

II. Cattle

A summary of the measurements taken on cattle limb bones from Brandon, Ipswich and Wicken Bonhunt is included in Table 5.1. The withers height estimates are included in Table 5.2. Following von den Driesch and Boessneck (1974, 336), Fock's factors were used to calculate withers heights for metapodia, while Matolcsi's factors were used for the other long bones. For comparative purposes, the tables also include the metrical data for cattle from the Iron Age settlement at West Stow

(Crabtree 1990a; 1990b), the 5th- and 6th-century contexts from the West Stow Anglo-Saxon village (Crabtree 1990a), and the late Roman settlement at Icklingham (Crabtree 1991; Crabtree 2010a). MacKinnon (2010a) has recently shown that Roman cattle in Italy showed substantial size improvement between the Republican and the Imperial Periods. A number of studies from different parts of the Empire have shown that the Romans introduced larger cattle to Europe and North Africa (see, for example, Teichert 1984; Albarella et al. 2008; MacKinnon 2010b). In addition, diachronic studies of cattle from the British Isles have shown that Iron Age cattle were small, that the Romans introduced larger cattle to Britain, and that the Anglo-Saxon cattle maintained some of the size improvement that began in Roman times (see, for example, Maltby 1981; Albarella et al. 2008).

The data from East Anglia generally confirm these observations. Withers heights for the Iron Age cattle from West Stow averaged only 107cm. The largest of the Iron Age cattle had a withers height of only 116cm. The Roman cattle from Icklingham, however, had an average withers height of 119cm. In other words, the average Roman cattle from late Roman (primarily 4th century) Icklingham were larger than the largest cattle from Iron Age West Stow.

The Anglo-Saxon cattle from West Suffolk are intermediate in size between the Iron Age and the Roman cattle. The West Stow cattle have an average estimated withers height of between 112 and 114cm, while the Brandon cattle had a withers height of slightly less than 114cm. As described in detail elsewhere (Crabtree and Campana n.d.), statistical comparisons between the West Stow 6th-century cattle and the cattle from Brandon reveal no evidence for size changes in cattle from West Suffolk from the Early to the Middle Saxon periods. The Ipswich and Wicken Bonhunt cattle are slightly larger with estimated withers heights of 117 and 118cm respectively.

A similar pattern is seen in the greatest lateral length (GLl) of the astragalus. The Iron Age sample from West Stow had the smallest astragali; the Icklingham cattle were largest, and the Anglo-Saxon cattle were intermediate in size. The Wicken Bonhunt cattle were slightly larger than those from the other Anglo-Saxon sites. The proximal breadth of the radius (Bp) and the trochlear breadth of the humerus (BT) show similar patterns. Since these are weight-bearing elements, they are likely to reflect the overall size of the cattle.

While summary statistics can reveal broad patterns of size change through time, distributional data can reveal some of the variation within individual assemblages. For the Middle Saxon faunal samples, one critical question concerns the roles that females and castrated males played in the economy. The Brandon assemblage includes the largest number of measured specimens and the most measurable metacarpals. Howard (1963) suggested that the metapodial indices (Bd × 100/GL and SD × 100/ GL) could be used to separate male, female, and castrated cattle. In theory, female cattle would be short and slender, while intact bulls would be far more robust. Since

Measurement	Mean	Min.	Max	s	C.V.	N
Humerus BT						
Brandon	69.5	60.4	92.1	5.4	7.8	37
Ipswich	67.2	58.8	81.3	4.7	6.9	52
Wicken Bonhunt	67.3	59.6	76.6	4.7	7.0	39
West Stow Iron Age	58.1	51.1	65.0			2
Icklingham	69.7	65.5	76.3	3.0	4.3	14
West Stow 5th century	66.7	63.0	68.3	2.1	3.1	5
West Stow 6th century	67.5	62.3	73.5	4.0	5.4	15
Radius GL						
Brandon	262.2	252.5	282.5	8.7	3.9	10
Wicken Bonhunt	269.7	245.3	298.2	17.3	6.4	16
West Stow Iron Age	246.2					1
West Stow 5th century	251.8					1
West Stow 6th century	259.9					1
Radius Bp						
Brandon	75.4	64.2	89.3	6.1	8.1	61
Ipswich	73.4	65.1	86.0	4.8	6.5	36
Wicken Bonhunt	76.0	65.9	86.7	6.5	8.5	50
West Stow Iron Age	73.5	71.0	76.5			2
Icklingham	78.2	67.6	92.5	7.4	9.5	10
West Stow 5th century	74.0	65.2	88.4	5.5	7.4	14
West Stow 6th century	74.1	62.8	86.0	5.6	7.6	25
Metacarpus GL						
Brandon	186.1	167.5	215.0	9.1	4.9	43
Ipswich	188.9	172.0	207.0	9.1	4.8	37
Wicken Bonhunt	194.8	180.6	213.0	9.7	5.0	10
West Stow Iron Age	180.1	166.7	189.8	10.4	5.8	5
Icklingham	193.3	181.5	209.0	6.6	3.4	13
West Stow 5th century	182.9	170.8	194.7	8.2	4.5	8
West Stow 6th century	187.0	176.9	198.2	7.8	4.2	13
Tibia Bd						
Brandon	57.0	50.2	67.2	4.1	7.2	147
Ipswich	57.8	49.5	69.3	4.8	8.3	92
Wicken Bonhunt	58.6	50.8	67.0	4.4	7.6	71
West Stow Iron Age	57.3	50.8	65.4	5.5	9.5	9
Icklingham	58.7	53.5	69.0	3.0	5.1	50
West Stow 5th century	55.9	50.8	67.4	4.0	7.2	23
West Stow 6th century	56.0	50.5	65.5	4.3	7.7	37
Astragalus GLl						
Brandon	61.1	53.8	69.7	3.2	5.2	269
Ipswich	61.5	53.3	72.4	3.9	6.3	108
Wicken Bonhunt	62.4	51.2	67.8	3.8	6.0	39
West Stow Iron Age	58.0	53.9	61.3	3.0	5.2	8
Icklingham	62.9	55.5	71.2	3.1	4.9	70
West Stow 5th century	61.1	54.2	65.8	3.2	5.2	27
West Stow 6th century	60.1	53.6	67.2	2.7	4.5	61
Metatarsus GL						
Brandon	210.0	191.0	244.5	13.9	6.6	14
Ipswich	216.3	204.5	229.7	8.5	3.9	18
Wicken Bonhunt	221.1	201.3	241.3	9.1	4.1	25
West Stow Iron Age	204.1	188.0	213.3			3
Icklingham	219.9	205.0	242.5	11.0	5.0	10
West Stow 5th century	207.2	192.6	221.8			2
West Stow 6th century	211.7	204.8	222.4			3
Metatarsus Bd						
Brandon	51.2	44.4	60.7	4.7	9.1	52
Ipswich	51.5	46.5	62.6	4.1	8.0	42
Wicken Bonhunt	52.6	44.9	60.3	4.7	8.9	31
West Stow Iron Age	54.2	53.2	56.2			3
Icklingham	51.3	46.7	58.9	2.4	4.7	62
West Stow 5th century	52.3	46.8	61.1	4.9	9.4	15
West Stow 6th century	49.1	44.9	58.6	3.1	6.3	31

Table 5.1 Summary statistics for measurements taken on cattle limb bones from Brandon, Ipswich and Wicken Bonhunt (all measurements in mm)

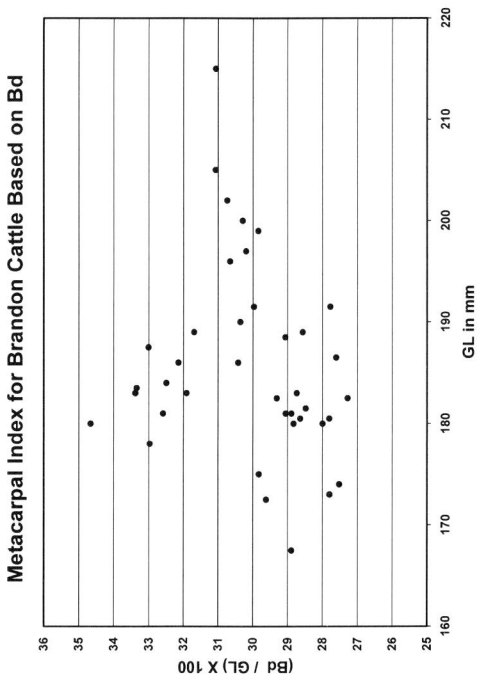

Figure 5.1 Metapodial Index (Bd/GL) × 100 for cattle metacarpals from Brandon

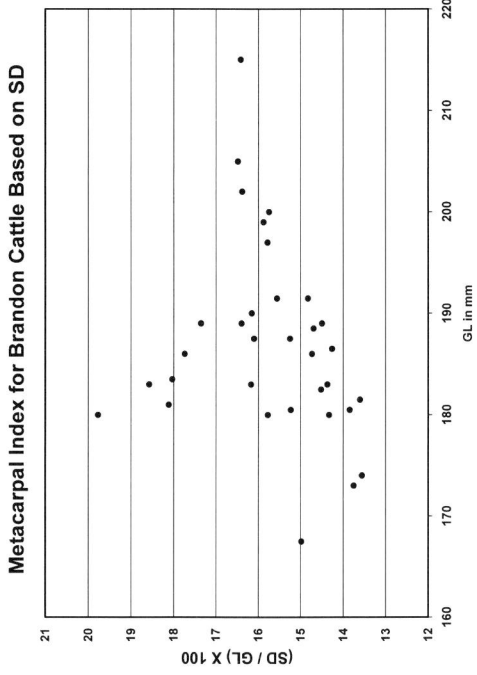

Figure 5.2 Metapodial Index (SD/GL) × 100 for cattle metacarpals from Brandon

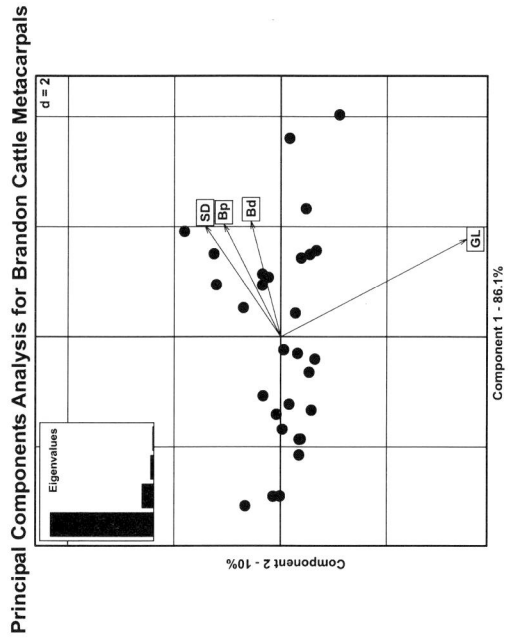

Figure 5.3 Graph showing the first two components of a Principal Components Analysis (PCA) for cattle metacarpals from Brandon (following Chessel *et al.* 2004)

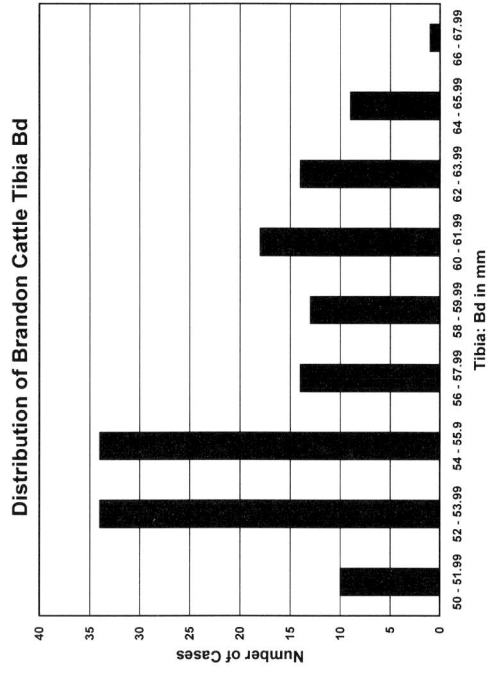

Figure 5.4 Distribution of distal tibial breadths (Bd) for cattle from Brandon

Metacarpal Index for Ipswich Cattle Based on Bd

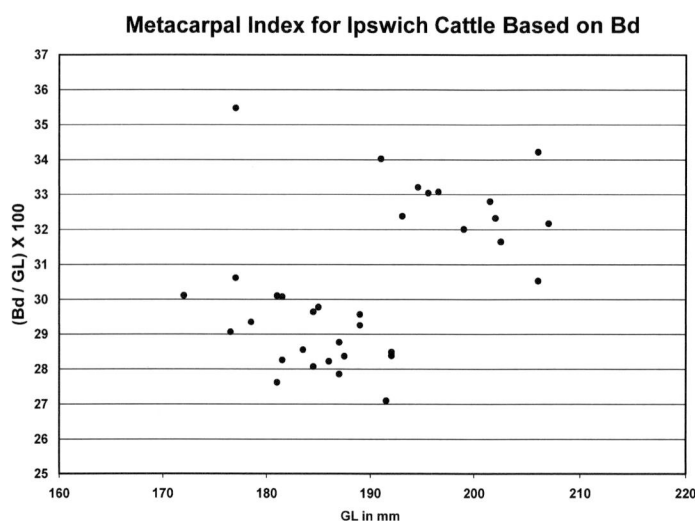

Figure 5.8 Metapodial Index (Bd/GL) × 100 for cattle metacarpals from Ipswich

animals that remained were used primarily for traction and transport purposes.

The Ipswich faunal assemblage, like the collections from the *emporia* of York, London and Hamwic, was dominated by the remains of cattle. The osteometric data may allow us to understand how this *emporium* was provisioned with food, since many of its inhabitants were likely engaged in non-agricultural activities such as crafts and pottery production. The summary measurements statistics (Tables 5.1 and 5.2) suggest that the Ipswich cattle were intermediate in size between the Brandon and the Wicken Bonhunt cattle with an average estimated withers height of 116.8cm.

Metapodial indices (Bd x 100/GL) were plotted against the greatest length of the metacarpus (GL) for the complete metacarpi from Ipswich (Figure 5.8). The data show two clusters: a shorter, more gracile group with metacarpal indices generally below 30, and a longer, more robust group with metacarpal indices above about 31. It seems reasonable to assume that these two clusters represent cows and steers, respectively. The single outlier

is a short, very robust specimen with a metacarpal index of more than 35. This probably represents an intact bull. These data suggest that the inhabitants of Ipswich were provisioned with beef from market-age and older adult cows and steers, and that the cows outnumbered the steers. There was little evidence for pathology on the post-cranial skeletons of these animals, so these are probably not worn-out traction animals.

The distribution of the distal tibial breadths (Bd) shows a similar bimodal pattern (Figure 5.9). Many of the cattle are relatively small, with distal tibial breadths ranging from 52 to 56mm, but a smaller group of tibiae is appreciably larger, with distal tibial breadths between 64 and 66mm. The Shapiro-Wilk test indicates that the probability of this distribution being drawn from a normal population is $p = 0.00007$.

The excavations at Ipswich also yielded substantial collection of measureable cattle from early Late Saxon, middle Late Saxon, and Early Medieval contexts. These data have been discussed in some detail elsewhere (Crabtree 2012a), but it is worth noting that the osteometric

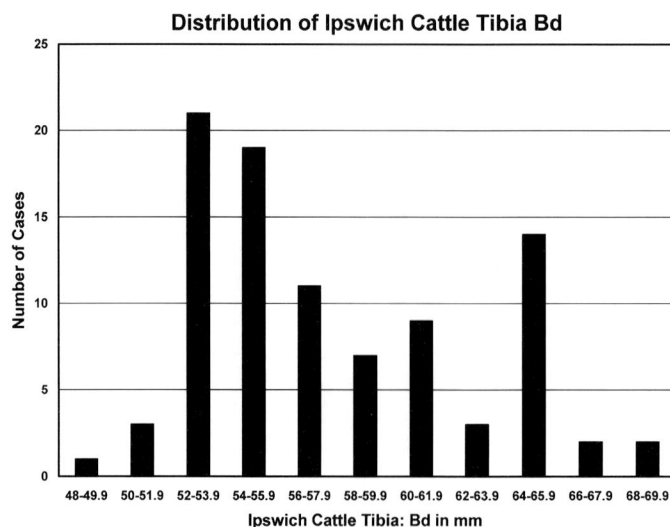

Distribution of Ipswich Cattle Tibia Bd

Figure 5.9 Distribution of distal tibial breadths (Bd) for cattle from Middle Saxon Ipswich

data suggest that the size of the cattle at Ipswich decreases from the Middle Saxon to Early Medieval periods. The average withers height for the Early Medieval cattle is only 112.3cm.

What conclusions can be drawn from the osteometric data for Middle Saxon cattle from East Anglia? The measurement data show that these Anglo-Saxon cattle are larger than Iron Age cattle, and they appear to maintain some of the size increase that was introduced by the Romans. Not surprisingly, the Middle Saxon cattle from Brandon are generally similar in size to the Early Saxon cattle from the neighbouring site of West Stow. The Middle Saxon cattle from Brandon, Ipswich and Wicken Bonhunt are generally similar in size. However, the cattle from Wicken Bonhunt are, on average, slightly larger than those from the other two sites. The metacarpal indices suggest that the Wicken Bonhunt assemblage is composed primarily of steers, while the Ipswich and the Brandon assemblages include more female cattle. It therefore seems reasonable to suggest that the size differences that are seen among the Middle Saxon cattle assemblages are likely to reflect differences in sex ratios rather than real differences in cattle sizes.

III. Sheep

Sheep have played a major role in East Anglian agriculture since Early Saxon times. The faunal assemblage from the West Stow village was dominated by the remains of sheep, and ageing data suggest that the sheep were kept for a variety of purposes, including meat, milk and wool. Faunal data from other Early Saxon sites and from contemporary Merovingian sites in northern France indicate the early medieval sheep husbandry was unspecialised (Crabtree 2010b). Ageing data from Brandon indicate that more specialised wool production was established in East Anglia during the Middle Saxon period (Chapter 4). Is this change reflected in the osteometric data?

The summary measurements for sheep from Brandon, Ipswich, and Wicken Bonhunt are listed in Table 5.3, and the withers height estimates are shown in Table 5.4. The withers height estimates are based on Teichert's factors following von den Driesch and Boessneck (1974). The data from Iron Age West Stow, the 5th- and 6th-century contexts at West Stow, and late Roman Icklingham are included for comparative purposes. The overall pattern here is different from the patterns seen in the cattle measurements. The data on the Iron Age sheep are relatively limited, but they tend to be small. The Roman sheep are large, but the Early Saxon sheep from West Stow appear to preserve much of the size improvement seen in the Roman sheep. The sheep from Icklingham have an average withers height of 62.6, while the West Stow sheep average 61.7–61.9cm. The Middle Saxon sheep are generally smaller than the Early Saxon sheep with average withers heights ranging between 56.6cm (Brandon) and 59.4cm (Wicken Bonhunt). The distribution of the estimated withers heights for the Brandon, Ipswich, Wicken Bonhunt and West Stow sheep is shown in Figure 5.10.

As described elsewhere (Crabtree 2007; Crabtree and Campana n.d.), a series of two-tailed t-tests were used to compare the Brandon sheep measurements to the measurements taken on sheep from the 6th-century

contexts at West Stow. The measurements compared include the humerus (BT), radius (Bp), metacarpus (Bp, Bd, GL), femur (Bp, DC), tibia (Bd), calcaneus (GL), astragalus (GLl), and metatarsus (Bp, Bd, GL). The t-tests shows the proximal breadths and greatest lengths of both the metacarpus and metatarsus of the sheep from Brandon were significantly smaller than their West Stow counterparts (p = .01). In addition, the distal breadth of the metatarsus and the greatest length of the calcaneus were significant at the p = .05 level. The proximal breadth of the sheep femora from Brandon, however, were significantly larger than those from West Stow at the p = .05 level. These data suggest that the Middle Saxon sheep from Brandon were generally smaller than the sheep from the Early Saxon village of West Stow, although the data from the femora suggest that pattern may be more complicated. If the proximal breadth of the femur is more closely correlated with sex than with overall body size, then the larger size of the Brandon femora may reflect the predominance of male sheep in the Brandon assemblage.

The other size differences cannot be easily attributed to changes in the sex ratio. The Brandon assemblage is made up primarily of (presumably larger) male sheep. Of the 521 horn cores and pelves from Brandon that could be sexed with reasonable certainly (following Tyler 1987 for the acetabula), 309 or approximately 59% were male. The high proportion of males makes sense if the Brandon farmers were interested in wool, since males, and particularly castrated males or wethers, are excellent wool producers.

The Middle Saxon sheep from Ipswich were small, with an average withers height of only 57.8cm. They are generally smaller than the early Late Saxon sheep from Ipswich, which had an average withers height of 60.3cm. Two-tailed Student's t-tests were used to compare the measurements taken on the Middle and early Late Saxon sheep from Ipswich. The differences between the radius Bp, metacarpus GL, tibia Bd, and metatarsus Bd were non-significant. The differences between the radius GL, metacarpus Bd, and astragalus GLl were significant at the p = .05 level, while the differences between the metatarsus GL and the estimated withers heights were significant at the p = .01 level. In all cases, the Middle Saxon sheep were significantly smaller than their early Late Saxon counterparts.

The Wicken Bonhunt sheep are somewhat larger, based on their average withers heights, than the sheep from either Brandon or Ipswich. The withers heights from Wicken Bonhunt reveal a distinctly bimodal distribution that is very different from the withers height distributions for the other three Anglo-Saxon sites (Fig. 5.10). If this distribution reflects sexual dimorphism, then the Wicken Bonhunt sheep assemblage includes a large number of females and a smaller number of male animals. This is very different from what we see at Brandon and might suggest that the Wicken Bonhunt sheep are part of a different population. However, the sample of complete sheep long bones from Wicken Bonhunt is relatively small.

The distribution of the distal tibial breadths (Bd) for sheep from Wicken Bonhunt is shown in Figure 5.11. These data appear to be more normally distributed, and the probability of these data being drawn from a normal population is 0.19, based on the Shapiro-Wilk test. When these data are compared to the tibial Bd (Fig. 5.12) from

Measurement	Mean	Min.	Max	s	C.V.	N
Humerus BT						
Brandon	28.1	24.3	32.7	1.5	5.3	390
Ipswich	28.3	23.4	34.9	2.1	7.6	118
Wicken Bonhunt	27.6	23.8	31.7	1.7	6.1	59
West Stow Iron Age	27.4	25.6	29.5	1.4	5.1	5
Icklingham	28.6	25.6	32.0	1.6	5.6	31
West Stow 5th century	28.3	25.1	32.0	1.7	6.0	31
West Stow 6th century	27.7	24.5	31.9	1.7	6.1	94
Radius GL						
Brandon	141.7	124.0	157.0	8.3	5.9	55
Ipswich	142.8	130.8	161.0	8.3	5.8	13
Wicken Bonhunt	146.9	126.4	158.4	9.9	6.7	11
Icklingham	146.0	131.0	161.0			2
West Stow 5th century	153.7	138.7	163.3	8.9	5.8	6
West Stow 6th century	146.0	135.3	161.3	8.0	5.5	10
Radius Bp						
Brandon	30.5	25.7	35.1	1.8	5.9	295
Ipswich	31.5	28.1	35.8	2.1	6.7	33
Wicken Bonhunt	29.9	25.7	33.4	1.7	5.8	62
West Stow Iron Age	30.9	29.3	33.0			3
Icklingham	30.1	27.1	35.6	1.9		18
West Stow 5th century	31.1	27.7	35.6	2.3	7.4	18
West Stow 6th century	30.2	25.1	35.9	2.4	7.9	44
Metacarpus GL						
Brandon	117.5	106.5	131.0	5.7	4.8	75
Ipswich	118.7	105.4	128.0	8.2	6.9	9
Wicken Bonhunt	120.7	113.1	129.6			4
West Stow Iron Age	127.6	125.1	130.1			2
Icklingham	124.3	109.9	134.0	8.4	6.1	6
West Stow 5th century	127.2	110.1	141.5	9.9	7.8	19
West Stow 6th century	128.7	117.1	142.1	7.3	5.6	18
Tibia Bd						
Brandon	25.8	19.5	29.5	1.3	5.0	735
Ipswich	26.3	22.4	29.7	1.5	5.7	88
Wicken Bonhunt	25.7	22.2	29.1	1.2	4.9	136
West Stow Iron Age	25.6	17.9	27.9	2.6	10.2	13
Icklingham	25.9	21.4	29.1	1.8	6.9	61
West Stow 5th century	26.2	22.4	27.9	1.7	6.5	42
West Stow 6th century	26.0	22.8	29.5	1.5	5.8	96
Astragalus GLl						
Brandon	27.5	23.6	31.3	1.3	4.2	247
Ipswich	27.7	25.7	30.4	1.1	4.0	19
Wicken Bonhunt	26.4	24.3	27.7			4
West Stow Iron Age	26.9	23.8	31.9			3
Icklingham	29.8	27.4	31.8	1.1	3.7	12
West Stow 5th century	28.0	26.0	29.9	1.4	5.0	25
West Stow 6th century	28.1	24.8	31.6	1.6	5.7	70
Metatarsus GL						
Brandon	126.0	110.5	143.5	7.6	6.0	44
Ipswich	128.0	119.3	139.0	5.7	4.5	14
Wicken Bonhunt	133.0	121.6	143.6	7.1	5.3	13
Icklingham	143.4	132.5	156.0	9.7	6.8	8
West Stow 5th century	134.7	122.2	148.2	8.0	5.9	7
West Stow 6th century	138.4	128.6	150.4	6.3	4.6	17
Metatarsus Bd						
Brandon	23.1	20.8	25.9	1.2	5.2	73
Ipswich	24.1	21.5	28.0	1.7	7.1	18
Wicken Bonhunt	23.6	21.9	25.0	1.0	4.3	19
West Stow Iron Age	22.4	22.3	22.5			2
Icklingham	24.5	21.4	27.5	2.2	9.0	10
West Stow 5th century	23.1	21.6	24.7	0.9	2.9	10
West Stow 6th century	24.0	21.1	26.4	1.3	5.4	30

Table 5.3 Measurements on sheep bones from Brandon, Ipswich and Wicken Bonhunt

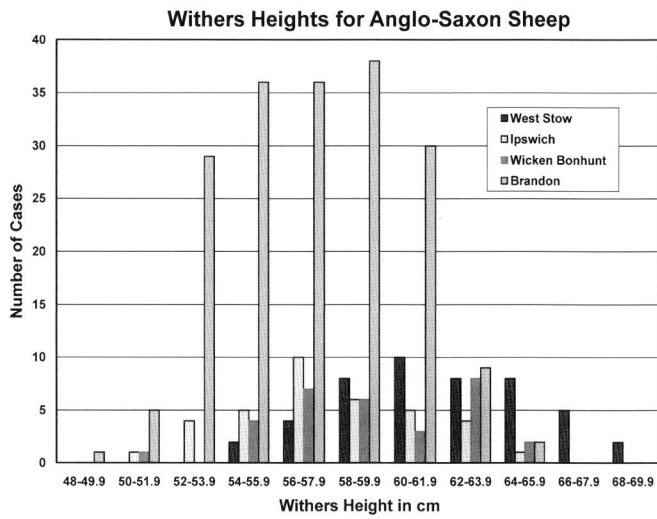

Figure 5.10 Distribution of withers heights for sheep from Brandon, Ipswich, Wicken Bonhunt and West Stow

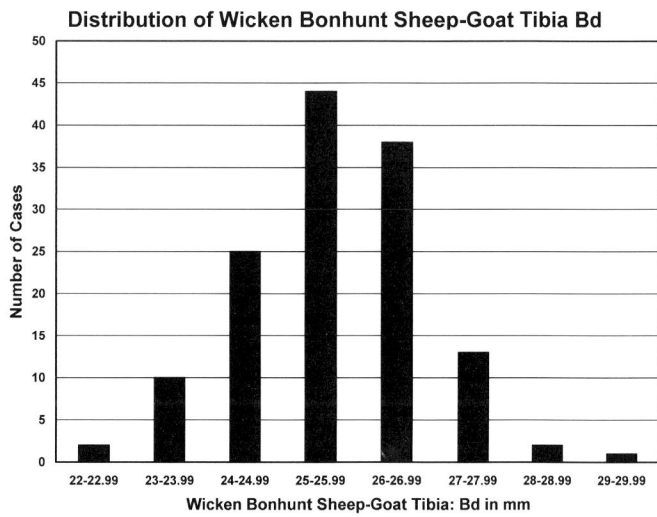

Figure 5.11 Distribution of distal tibial breadths (Bd) for sheep-goat from Wicken Bonhunt

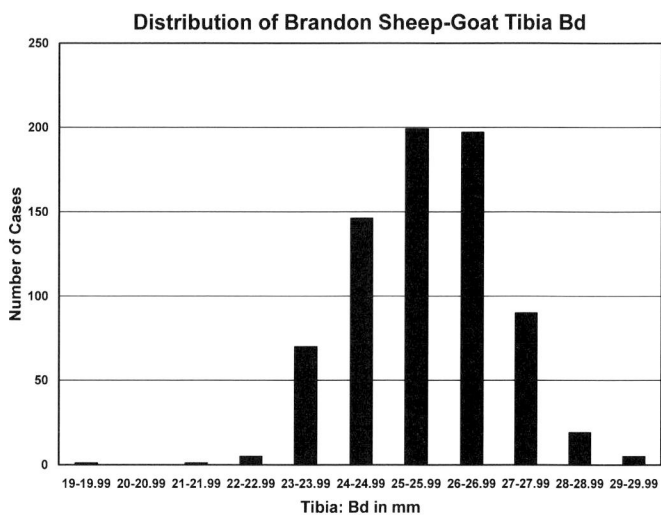

Figure 5.12 Distribution of distal tibial breadths (Bd) for sheep-goat from Brandon

51

	Mean	Min	Max	N
Brandon	56.6	49.8	65.1	186
Ipswich	57.8	51.5	64.7	36
Wicken Bonhunt	59.4	50.8	65.2	31
Icklingham	62.6	52.7	70.8	16
West Stow 5th century	61.7	53.8	68.6	34
West Stow 6th century	61.9	54.4	69.5	47

Table 5.4 Estimated withers heights in cm for sheep from Brandon, Ipswich and Wicken Bonhunt

Brandon using a two-tailed Student's t-test, the differences are non-significant (p =.63). These data might suggest that differences in bone length, as seen in the estimated withers heights, are not correlated with differences in bone breadth for the Middle Saxon sheep from East Anglia.

IV. Pigs

Even though the faunal assemblages from Brandon, Wicken Bonhunt and Ipswich are relatively large, the numbers of measured pig bones are fairly small (Table 5.5). Although the Wicken Bonhunt assemblage was dominated by the remains of pigs, most of the bones were cranial elements and teeth. Relatively few measureable post-cranial elements were recovered. There is no clear evidence for the presence of wild boar at any of these sites; all the dental measurements are well within the range for domesticated pigs. It seems reasonable to suggest that boar was rare in East Anglia during Middle Saxon times. The complete limb bone measurements and their associated withers heights (based on Teichert's factors following von den Driesch and Boessneck (1974)) are shown in Table 5.6. The Wicken Bonhunt assemblage included five complete long bones with estimated withers heights ranging from 65.1 to 77.9cm. The two withers height estimates for the Brandon pigs are 71.1 and 74.7cm. Two complete limb bones from Ipswich provided withers heights of 69.2 and 76.9cm.

The largest number of Middle Saxon pig measurements comes from Brandon. Two-tailed Student's t-tests were used to compare the following limb bone measurements to the measurements taken on pig bones from 6th-century West Stow: humerus Bd, radius Bp, tibia Bd, and astragalus GLl. The bone measurements

Measurement	Mean	Min.	Max	s	C.V.	N
Humerus Bd						
Brandon	37.0	33.4	41.8	1.9	5.1	107
Ipswich	37.8	26.7	41.5	3.3	8.7	23
Wicken Bonhunt	37.3	34.3	43.1	1.7	4.6	55
Icklingham	39.0	36.9	42.4	1.9	4.9	6
West Stow 5th century	38.6	34.1	43.6	2.4	6.2	12
West Stow 6th century	40.0	36.0	43.1	2.1	5.3	15
Radius Bp						
Brandon	27.4	23.7	32.3	1.6	5.8	134
Ipswich	28.1	25.0	31.7	1.6	5.7	67
Wicken Bonhunt	28.6	24.8	34.7	2.4	8.4	17
West Stow Iron Age	25.3					1
Icklingham	29.4	26.4	34.4	3.0	10.2	5
West Stow 5th century	28.6	25.5	31.0	1.5	5.2	10
West Stow 6th century	28.7	26.4	30.8	1.5	5.2	12
Tibia Bd						
Brandon	28.7	25.2	33.7	1.5	5.2	120
Ipswich	29.0	26.6	32.9	1.5	5.2	31
Wicken Bonhunt	33.7	26.3	38.9	1.8	6.2	17
Icklingham	39.4	32.0	46.7			2
West Stow 5th century	29.6	27.6	31.7	1.6	5.4	10
West Stow 6th century	29.4	27.2	31.0	1.2	4.1	11
Astragalus GLl						
Brandon	38.8	35.0	43.6	1.8	4.6	89
Ipswich	39.3	34.2	42.0	1.9	4.8	24
Wicken Bonhunt	35.3	31.7	40.7	2.1	5.4	6
Icklingham	38.3					1
West Stow 5th century	37.7	35.7	40.8	1.8	7.4	9
West Stow 6th century	39.5	33.9	45.0	2.5	6.3	14
LM3 Length						
Brandon	32.0	25.8	38.0	2.5	7.8	95
Ipswich	31.8	28.3	34.1	1.5	4.8	28
Wicken Bonhunt	31.0	25.1	37.2	2.1	6.7	418
West Stow Iron Age	33.0	28.4	35.1	2.7	8.2	5
Icklingham	34.7	31.6	38.6	2.8	8.1	5
West Stow 5th century	32.9	30.0	37.0	2.2	6.7	10
West Stow 6th century	31.8	29.9	35.1	1.4	4.4	11

Table 5.5 Measurements in mm on pig bones from Brandon, Ipswich and Wicken Bonhunt

on the Brandon pigs were consistently smaller than those from West Stow. The differences between the measurements on the radii and humeri from the two sites were significant at the p = .01 level. These data suggest that the size of pigs may have decreased in West Suffolk between the Early and the Middle Saxon periods. The Ipswich and Wicken Bonhunt pigs are generally similar in size to the Brandon pigs.

V. Horses

Horses played a variety of different roles in Anglo-Saxon economy, society, and ritual. As described in Chapter 2, butchery marks indicate that horses formed a part of the Early Saxon diet, at least on an occasional basis, at Early Saxon West Stow. Horse bones were also used as raw materials for bone working (Crabtree 1990a, 104). Horses often accompanied high-status pagan Anglo-Saxon burials, as seen at sites such as Snape, Sutton Hoo, and Eriswell (Filmer-Sankey and Pestell 2001, 256, see also Fern 2007) in Suffolk. As noted above (Chapter 3), a horse appears to have been used as a foundation deposit at Middle Saxon Brandon. Osteometric analyses showed that most of the Early Saxon horses from West Stow were, in fact, large ponies about the size of a modern New Forest pony (Crabtree 1990a, 62). How do the horse remains from the Middle Saxon sites in East Anglia compare to these Early Saxon animals?

The measurements taken on the horse bones from Brandon have been summarised elsewhere (Crabtree and Campana n.d.). The withers height estimates for the complete horse long bones (based on Kieswalter's factors following von den Driesch and Boessneck (1974)) from Brandon, Ipswich and Wicken Bonhunt are summarised in Table 5.7. The Brandon horses have a mean withers height of 140.2cm, or just under 14 hands. They are significantly larger than Early Saxon horses from West Stow which had a mean withers height of 132.3cm (s.d. = 6.8, N = 15). A Student's two-tailed t-test reveals that the two samples are significantly different at the p = .01 level.

Only four complete horse long bones were recovered from the Middle Saxon contexts at Ipswich. The estimated withers heights for these animals range from 132.3cm to 139.3cm, with a mean of 134.6cm. While the sample is too small for statistical analysis, the Ipswich horses seem to be broadly similar in size to the Early Saxon horses from West Stow.

The horse remains from Wicken Bonhunt present a somewhat different picture. Two complete long bones yielded withers height estimates of 129.7 and 159.4cm. The former is a pony of just under 13 hands; the latter is a large horse measuring over 15.2 hands that is atypical of Anglo-Saxon horses in eastern England. It is appreciably larger than the horse remains recovered from West Stow, Brandon and Ipswich, and it is also larger than the horse burial from late Roman Icklingham (Levine *et al.* 2002).

A small number of horse bones from both Brandon and Ipswich showed evidence for chopping and splitting, suggesting that horses were at least an occasional part of the Middle Saxon diet in East Anglia. As noted in Chapter 3, a more extensive series of butchery marks was observed on 59 horse limb bone elements from Middle Saxon Wicken Bonhunt. These examples include several cases of the longitudinal splitting of the long bones, presumably for marrow extraction. At Brandon, cattle long bones were

Site	Limb Bone	GL (mm)	WH (cm)
Brandon	Radius	142.0	74.7
Brandon	Ulna	179.0	71.1
Ipswich	Radius	146.2	76.9
Ipswich	Tibia	176.5	69.2
Wicken Bonhunt	Radius	138.3	72.7
Wicken Bonhunt	Radius	123.7	65.1
Wicken Bonhunt	Radius	131.2	69.0
Wicken Bonhunt	Radius	148.1	77.9
Wicken Bonhunt	Tibia	171.9	67.4

Table 5.6 Estimated withers heights (cm) for complete pig long bones from Brandon, Ipswich and Wicken Bonhunt

frequently split for the extraction of marrow (Crabtree and Campana n.d.). The presence of these butchery traces on the horse bones from Wicken Bonhunt suggests that these horses may have formed part of the Middle Saxon diet at the site. There is no pathological evidence to suggest that the Bonhunt horses were used for traction purposes, but four fused thoracic vertebrae may indicate the presence of spondylitis which results from repeated back injuries and can be caused by breaking a horse for riding or by extensive riding.

VI. Dogs

The West Stow Anglo-Saxon village produced several dog skeletons, as well as a number of other dog limb bone elements. The West Stow dogs were generally large, straight limbed animals, with an average withers height of 59.5cm. A Roman dog skeleton recovered from the pottery kilns at the site was smaller, with an estimated withers height of only about 40cm (Crabtree 1990, 65).

The five complete dog bones from Brandon (Table 5.8) yielded an average withers height of 58.5cm (based on Koudelka's factors, following on den Driesch and Boessneck (1974)). Like the West Stow dogs, they are large and straight limbed. They would have been ideally

Site	Limb Bone	Ll (mm)	WH (cm)
Brandon	Tibia	330.0	143.9
Brandon	Tibia	330.0	143.9
Brandon	Radius	347.0	150.6
Brandon	Metatarsus	263.0	140.2
Brandon	Metatarsus	261.0	139.1
Brandon	Metatarsus	246.5	131.4
Brandon	Metacarpus	216.5	138.8
Brandon	Metacarpus	205.5	131.7
Brandon	Metacarpus	221.5	142.0
Brandon	Metacarpus	219.5	140.7
Ipswich	Radius	324.9	134.0
Ipswich	Metatarsus	252.9	132.3
Ipswich	Metatarsus	266.4	139.3
Ipswich	Metacarpus	214.9	133.3
Wicken Bonhunt	Tibia	297.5	129.7
Wicken Bonhunt	Metacarpus	248.6	159.4

Table 5.7 Withers height estimates (in cm) for complete horse limb bones from Middle Saxon Brandon, Ipswich and Wicken Bonhunt

Site	Limb Bone	GL (mm)	WH (cm)
Brandon	Femur	199.0	59.9
Brandon	Tibia	179.5	52.4
Brandon	Humerus	188.0	63.4
Brandon	Humerus	176.0	59.3
Brandon	Radius	180.5	58.1
Ipswich EL Saxon	Humerus	140.9	47.5
Ipswich EL Saxon	Humerus*	98.6	33.2
Ipswich EL Saxon	Radius	165.4	53.3
Ipswich EL Saxon	Radius	175.4	56.5
Ipswich EL Saxon	Radius*	87.9	28.3
Ipswich EL Saxon	Ulna*	105.7	28.2
Ipswich EL Saxon	Femur*	106.7	32.1
Ipswich EL Saxon	Tibia*	96.4	28.1

Table 5.8 Withers height estimates for Middle Saxon dogs from Brandon and early Late Saxon dogs from Ipswich. The elements marked with an asterisk come from a single small dog from Ipswich

suited to tasks such as hunting and guarding (Harcourt 1974, 168). Few dog bones were recovered from Wicken Bonhunt, and no measurable post-cranial dog bones were recovered, so we can say almost nothing about the nature of the dogs at Wicken Bonhunt. No complete dog bones were recovered from the Middle Saxon contexts at Ipswich, but several dogs were recovered from early Late Saxon features (Table 5.8). Unlike the Brandon and West Stow dogs, these urban dogs show a great deal of variability. A number of bones were recovered from a single small dog that would have had a withers height of about 30cm. The other Ipswich dogs are larger with withers heights between 47.5 and 56.6cm. The largest of these are similar in size to the Brandon and West Stow dogs. The Ipswich dogs may have served a variety of different purposes, from guard dogs to pampered pets.

VII. Chicken

Domestic fowl were an important part of the Middle Saxon animal economy. In addition to eggs, these animals provided meat and feathers. Table 5.9 summarises the measurements taken on chicken bones from Brandon, Ipswich and Wicken Bonhunt. The greatest lengths (GL) of the chicken limb bones from Brandon show a distinctly bimodal distribution (Crabtree and Campana n.d., fig. 10). Does this represent sexual dimorphism, or does this represent two different breeds of chickens?

In order to answer this question, the greatest lengths (GL) of the chicken tarsometatarsals with and without measurable spurs from Ipswich and Wicken Bonhunt were compared (Fig. 5.13). The chickens with spurs are likely to be male. Serjeantson (2009, 48) notes that small spurs are present on some modern breeds of hens, but she suggests that the incidence of hens was spurs was likely to have been lower in the past. Those without spurs are likely to be female, but they also may include some capons and males whose spurs were removed or did not form (depending on the age of caponisation). Despite these cautions, Serjeantson (2009, 48) notes that the presence of a spur or spur scar is the best guide to the sex ratio of adult chickens. Since most of the chickens from these sites are mature, it seems reasonable to use the presence or absence of a spur or spur scar to distinguish males from females. The data in Figure 5.13 clearly suggest that the males are, on the whole, appreciably bigger than the females. From these data it might be reasonable to conclude that the Middle Saxon flocks were made up of large numbers of hens and much smaller numbers of roosters and capons. These Anglo-Saxon chickens may have been used primarily for egg-laying, although we did not specifically look for medullary bone at the time the bones were originally identified.

Measurements of the greatest length (GL) of the femur and humerus were converted to z-scores for the Brandon, Wicken Bonhunt and Ipswich fowl (Figs 5.14–5.16). All

Measurement	Mean	Min.	Max	s	C.V.	N
Humerus GL						
Brandon	68.6	59.9	82.6	5.1	7.4	66
Ipswich	68.7	61.6	81.2	5.3	7.7	31
Wicken Bonhunt	67.0	59.0	78.5	4.7	7.0	199
Ulna GL						
Brandon	68.1	57.7	74.9	5.0	7.3	41
Ipswich	67.1	57.4	79.0	6.0	8.9	46
Wicken Bonhunt	65.9	57.7	76.6	4.9	7.5	68
Femur GL						
Brandon	75.5	66.0	85.3	5.9	7.8	84
Ipswich	73.6	56.2	84.4	5.9	8.0	30
Wicken Bonhunt	73.0	63.4	82.9	5.1	7.0	25
Tibiotarsus GL						
Brandon	103.7	93.1	118.4	9.3	9.0	13
Ipswich	103.5	89.0	117.6	8.8	8.5	108
Wicken Bonhunt	100.5	91.3	116.7	7.1	7.1	21
Tarsometatarsus GL						
Brandon	74.3	64.0	88.1	10.3	13.9	5
Ipswich	69.4	56.0	84.1	7.5	10.8	44
Wicken Bonhunt	68.8	61.5	84.6	5.6	8.2	70

Table 5.9 Measurements in mm on chicken bones from Middle Saxon Brandon, Ipswich and Wicken Bonhunt

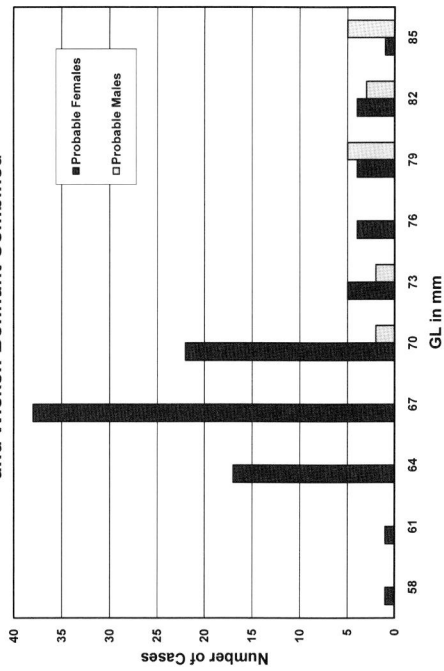

Figure 5.14 Distribution of z-scores of the greatest length (GL) of the chicken femora and humeri from Brandon

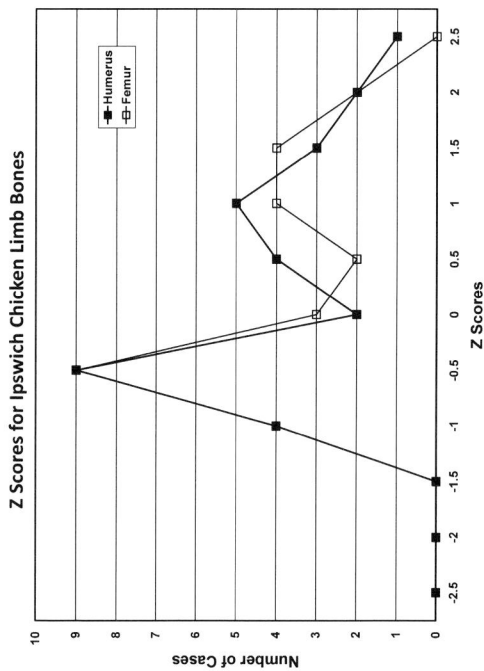

Figure 5.16 Distribution of z-scores of the greatest length (GL) of the chicken femora and humeri from Wicken Bonhunt

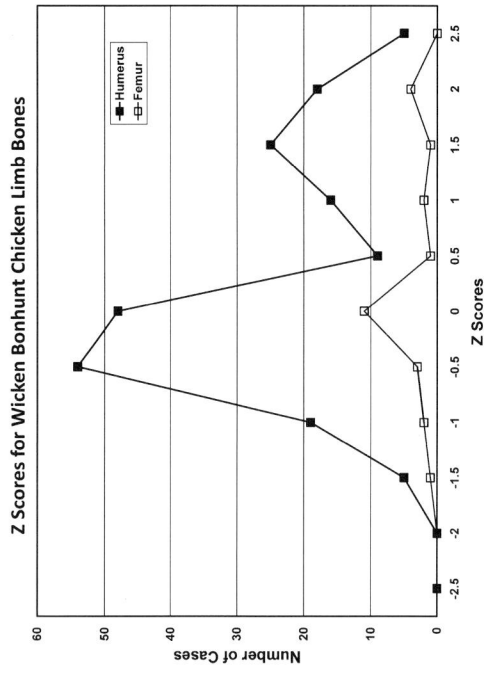

Figure 5.13 Distribution of the greatest length (GL) of the tibiotarsus for chickens from Ipswich and Wicken Bonhunt (those labelled male had a spur; those labelled female did not)

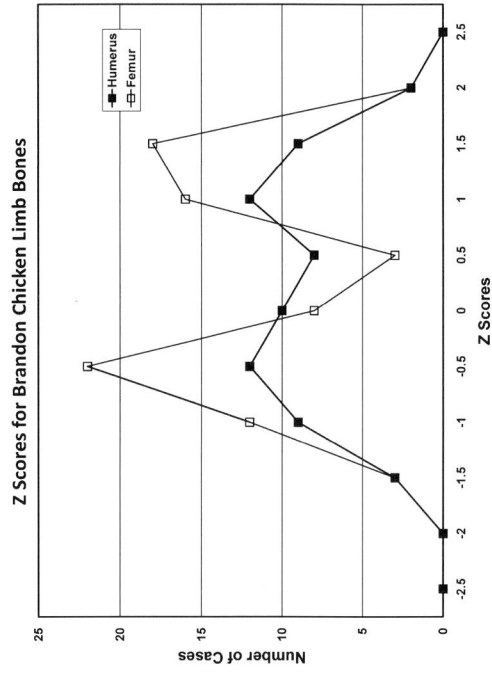

Figure 5.15 Distribution of z-scores of the greatest length (GL) of the chicken femora and humeri from Ipswich

three distributions show substantial bimodality which may reflect sexual dimorphism.

VIII. Summary

Comparisons of the Middle Saxon metrical data with the Iron Age, Roman and Early Saxon bone measurements from East Anglia has allowed us to trace changes in animal size through time. What is most interesting is that the different species reveal somewhat different patterns of size change. The cattle show the most predictable patterns of change. The Iron Age cattle from West Stow are small; the Roman cattle from nearby Icklingham are significantly larger; and the Early and Middle Saxon cattle preserve some of the size increase that was introduced by the Romans. The differences in cattle size amongst the three Middle Saxon assemblages seem to reflect differences in the sex ratios rather than significant differences in cattle size during the Middle Saxon period. Data from Ipswich indicate that cattle sizes decreased from the Middle Saxon to Late Saxon and Early Medieval periods (Crabtree 2011).

The sheep present a somewhat different picture. As is the case with the cattle, Iron Age sheep are relatively small, and the Roman sheep are appreciably larger. The Early Saxon sheep appear to retain most of the size increase that was introduced by the Romans. The Middle Saxon sheep are generally smaller than the Early Saxon ones. In particular, the Middle Saxon sheep from Brandon are significantly smaller than the Early Saxon sheep from West Stow. This does not seem to be a result of changing sex ratios, since the majority of the sheep from Brandon are, in fact, males.

The metrical data for horses and dogs indicate that size variation increased during the Middle Saxon and early Late Saxon periods. Data from the Roman features at West Stow and elsewhere in Roman Britain indicate that Roman dogs were variable in size, including both large animals and toy and dwarf breeds (see Baxter 2010a for an up-to-date discussion of small Roman dogs). The Early Saxon dogs from West Stow and the Middle Saxon dogs from Brandon, however, were large, straight-limbed animals, with withers heights just under 60cm. They are comparable in size to a modern Alsatian (German shepherd). The early Late Saxon dogs from Ipswich show much greater variation, including both larger dogs and a small dog, about the size of a terrier, with a withers height of about 30cm.

The metrical data for horses also point to increasing variability in animal size beginning in the Middle Saxon period. Early Saxon horses are generally small, about the size of a modern New Forest pony. The Middle Saxon horses from Brandon are, on average, slightly larger, with an average withers height of about 140cm. The data from Wicken Bonhunt, however, point to the existence of some larger horses, nearly 160cm in withers height, during the Middle Saxon period. The role of these larger horses remains unknown. Were they bred for use in warfare, for traction, or for some other purpose? Metrical analysis of other large Middle Saxon faunal samples may allow us to answer this question more fully.

Chapter 6. Discussion and Conclusions

I. Introduction

In this final chapter we will explore the kinds of information that quantitative analyses, mortality profiles, and osteometric data can provide about Middle Saxon animal husbandry practices and hunting patterns in East Anglia and in eastern England. The chapter will begin with an updated perspective on Early Saxon animal husbandry and the transition from the Early Saxon to the Middle Saxon period. The zooarchaeological data from Brandon, Ipswich, Wicken Bonhunt and other contemporary sites will then be used to examine the varied nature of Middle Saxon animal exploitation in eastern England.

II. An Updated Perspective on the Transition from Early to Middle Saxon England

When the faunal remains from the Early Saxon village of West Stow were first published over 20 years ago, the West Stow assemblage was the only large Early Saxon faunal collection that had been analysed using modern archaeozoological methods. Since then, a number of other Early Saxon faunal collections have been analysed, and comparison of these assemblages with the West Stow collection allows us to draw some broader conclusions about Early Saxon animal husbandry practices.

Osteometric data from West Stow indicate that the animals, and particularly the sheep, preserved some of the size increase that was introduced to Britain by the Romans. This is not universally the case for Early Saxon faunal assemblages from eastern England. In particular, the sheep from Kilham in Yorkshire had an average withers height of only 56.3cm (Archer 2003, table 13). A very small sample of sheep from the Early to Middle Saxon site of Bloodmoor Hill in Suffolk has an average withers height of only 54.9cm (Higbee 2009, table 23). It seems reasonable to suggest that one of the reasons that the inhabitants of West Stow may have had access to larger animals is the site's proximity to the late Roman site of Icklingham. While it is likely that Icklingham was abandoned before the settlement at West Stow was established, the inhabitants of both sites may have had access to closely related herds of livestock.

When I first analysed the faunal remains from West Stow, I concluded that the Early Saxons used improved animals in traditional ways (Crabtree 1982). Certainly the osteometric data from West Stow and the East Anglian Middle Saxon sites indicate that cattle preserved some of the animal size increase introduced by the Romans. However, when the West Stow mortality profiles for cattle and sheep are compared to the increasing body of evidence from other Early Saxon sites, the pattern of animal husbandry might better be described as unfocused (Rackham 2003). The mortality profiles for cattle and sheep do not reflect a focus on a single primary or secondary product, like wool, milk, meat or traction.

Cattle and sheep were probably raised and used for a variety of different purposes.

The kill-patterns and species ratios for the Early Saxon sites from eastern England suggest that these villages practiced a degree of autarky or economic self-sufficiency (*cf.* Yvenic and Chaulet 2009). This, however, should not be taken to mean that these villages were economically isolated. The recovery of the remains of a marine flatfish from West Stow clearly indicates a degree of contact with the coast. The survey work carried out on both sides of the English Channel by Loveluck and Tys (2006) points to an increasing cross-channel trade beginning as early as AD 600. It is also likely that the inhabitants of West Stow and other Early Saxon villages paid some form of taxes or tribute, probably in the form of food renders. Unfortunately, evidence for this is archaeologically invisible. It is a mistake to view each Anglo-Saxon village as an isolated entity; instead, it can be argued that much of the agricultural and pastoral produce was designed to meet local needs. However this pattern begins to change by the end of the Early Saxon period.

One of the most valuable contributions made by the excavators of the Anglo-Saxon settlement at Flixborough in Lincolnshire (Loveluck 2007; Dobney *et al.* 2007) was the revelation that social and economic changes do not necessarily correspond to the traditional chronological divisions within the Anglo-Saxon period. Careful analysis of the artefactual, structural, and faunal remains from this rural estate centre revealed two very different patterns of animal use and ways of life within the Middle Saxon period itself. The earlier occupation was secular and was characterised by lavish feasting. The estate appears to have fallen into ecclesiastical hands during the Middle Saxon period, and the later Middle Saxon period is characterised by increasing craft specialisation and a more austere lifestyle.

Some of the East Anglian estate centres begin to show a movement away from unfocused animal production as early as the later part of the Early Saxon period. While the data are limited, they are suggestive. Although the bulk of the faunal remains from Wicken Bonhunt are dated to the Middle Saxon period, a small faunal collection was recovered from three ditches and two pits which contained hand-made, sand-tempered pottery. These features have been dated to the late 6th-7th century and are therefore contemporary with the final phase of occupation at West Stow. These features yielded 2108 animal bones and fragments. The species ratios for the large domestic mammals indicate that the assemblage was dominated by pigs (61%), followed by cattle (21%), caprines (17%), and horses (1%). While the faunal assemblage from Early Saxon Wicken Bonhunt is small, the species ratios suggest that more specialised pork production may have begun at the site as early as the later 6th or early 7th century.

The recently published faunal data from Bloodmoor Hill (Higbee 2009) in eastern Suffolk are also suggestive. The site yielded the remains of a cemetery and a settlement that may represent an early estate centre. The faunal remains from the site are dominated by the remains of domestic mammals, including cattle, pigs and caprines. The mortality profiles for cattle and sheep differ significantly from those seen at West Stow. Higbee (2009, 304) concludes that

> ...the settlement acted as a consumer of beef from herds kept primarily for traction and dairying. By contrast, the mortality profile for sheep indicates that sheep were locally reared and that wool-production was important to the settlement's economy....The most basic but nevertheless important aspect of the economy is self-sufficiency in pork production. This complex economic pattern is significantly different from contemporary rural sites such as West Stow and suggests that the move towards specialisation seen at later Middle Saxon rural sites *may have had its roots in the early Anglo-Saxon period* (emphasis added).

The data from Bloodmoor Hill and Wicken Bonhunt suggest that some Anglo-Saxon farmers were moving away from an unfocused strategy for animal production by the later part of the Early Saxon period.

III. Middle Saxon Animal Husbandry: Brandon, Wicken Bonhunt and Ipswich in Context

The three Middle Saxon sites discussed in detail here include two wealthy estate centres and an *emporium* or '*wic*'. These sites may not be typical of all Middle Saxon settlements in East Anglia. Smaller farming communities, like the Early to Middle Saxon settlement at Brandon Road in Thetford (Baxter 2010b), may have had different patterns of animal production and use. In particular, they may have provided taxes and tribute, in the form of food renders, to Middle Saxon royal sites and estate centres. We need a wide range of settlements, representing both different social classes and different environmental zones, in order to understand the full dimensions of animal use in Middle Saxon East Anglia. However, the detailed analysis of the faunal data from these three important sites can shed light on the ways in which the Middle Saxons produced and consumed meat and other animal products.

The most striking feature of the Brandon and Wicken Bonhunt faunal assemblages is how very different they are. Brandon is located in the Breckland regions of western Suffolk, and its faunal assemblage is dominated by the remains of sheep. Wicken Bonhunt, on the other hand, is located in an area where pannage for pigs was plentiful throughout the Anglo-Saxon era. However, the differences between the sites are far more than simply environmental ones. The Brandon sheep mortality profiles show a preference for mature and elderly males, many slaughtered between 4 and 8 years of age. While sheep produce their finest wool before three years of age, they will continue to produce wool for several more years before the quality and quantity declines. O'Connor (2010, 12) has recently argued that this will lead to a culling of 5–7 year olds, a pattern that is mirrored in the Brandon data. The Wicken Bonhunt faunal assemblage is dominated by the remains of cranial elements from older pigs, many culled around 3 years of age. As argued elsewhere (Crabtree 2010b), these data suggest that animal production was becoming increasingly focused and specialised, at least at some of the Middle Saxon estate centres.

There is evidence that several other estate centres were also loci for specialised production of food products and other goods. As noted above, Flixborough in Lincolnshire seems to have developed into a centre for wool and textile production in the later part of the Middle Saxon period. Ramsbury in Wiltshire (Haslam 1981) was a centre for iron production, and a food rent collection centre has been identified at the estate centre of Higham Ferrers in Northamptonshire (Hardy *et al.* 2007). At Higham Ferrers, a large complex of buildings and enclosures dating to the 8th century includes a large malting oven. The authors suggest that a large variety of produce came into this site and was then redistributed. These data suggest that many of these high status estate centres may have been engaged in complex webs of production and redistribution. Moreover, the archaeological and zoo-archaeological data indicate that animals and animal products, agricultural produce and manufactured items were moving around amongst Middle Saxon settlements as a result of both trade and tribute.

Another striking feature of the Wicken Bonhunt assemblage is the presence of a number of male cattle, including some with evidence for traction pathologies. Middle Saxon cattle clearly played an important role in both traction and transport. One possibility is that traction animals became more important as the agriculture expanded and the amount of land under cultivation increased. In that context, it is interesting to note that the open field system may have its origins in the 8th century (Oosthuizen 2007) and that an iron coulter has recently been recovered from a 7th-century context at the site of Lyminge in Kent.

The use of cattle as traction animals in the Middle Saxon period is accompanied by the increasing importance of sheep at many Anglo-Saxon sites. For example, the relative importance of sheep increases from the Early to the Middle Saxon periods at Quarrington in Lincolnshire (Rackham 2003) and at West Heslerton in Yorkshire (Dobney *et al.* 2007, 223). Sheep increase in importance within the Middle Saxon period at Flixborough in Lincolnshire (Dobney *et al.* 2007, 223), and there is an increase in the proportions of sheep between Early Saxon West Stow and Middle Saxon Brandon. Similar changes are apparent between Merovingian and Carolingian France (Yvenic and Chaulet 2009). Two processes are probably at work here. First, wool production is likely to have played an increasingly important role in the economies of Middle Saxon England and Carolingian France. Second, the role of cattle as symbols of wealth and power may have declined. Their role became more strictly economic in nature.

One of the most striking features of the Middle Saxon period, and of the 7th to 10th centuries in coastal northern Europe, is the appearance of the *emporia*. These towns — which include York, London, Hamwic and Ipswich in Britain and sites such as Quentovic in France, Ribe in Denmark and Birka in Sweden on the European continent — were centres of both regional and international trade and craft production. Since many of the inhabitants of Middle Saxon Ipswich were likely engaged in non-agricultural activities such as pottery production, the question of how these craftsworkers were supplied with meat and other food products is an important one.

The faunal assemblages from Middle Saxon Ipswich, like the contemporary assemblages from Anglian York, Hamwic, and Middle Saxon London, are dominated by the remains of cattle, including both market-age and elderly individuals. These must have been provided by farmers from the surrounding countryside in eastern Suffolk, but we have only limited archaeological evidence for these sites. The inhabitants of Middle Saxon Ipswich were provided with meat from a limited range of animal species; the Ipswich assemblage lacks the diverse range of wild birds that are seen at the rural East Anglian sites. The presence of a number of neonatal pigs and young pigs who were slaughtered in their first year of life suggests that pigs were being raised in and around the town itself. If this is the case, then the provisioning of Ipswich may not have had a dramatic impact on the surrounding countryside. Ipswich certainly provided a market for surplus cattle and sheep, but the establishment of Ipswich did not necessarily lead to a wholesale restructuring of the patterns of animal production in the neighbouring farming villages.

This raises the important question about the relationship between the rise of the *emporia* and the other social, political and economic changes that characterise the 'long 8th century'. Hodges (1982) argued that there was a close relationship between the establishment and growth of the *emporia* and state formation in England and on the continent. It is reasonable to suggest that Ipswich was founded and grew under the aegis of the East Anglian royal house. However, it is not clear whether the foundation of the *emporium* at Ipswich was a cause or an effect of many of the other social, political and economic changes that took place during the 7th and 8th centuries. In 1996, it seemed to be the foundation of the *emporia* that led to fundamental changes in the Middle Saxon countryside (Crabtree 1996), but the situation appears different now (Crabtree 2010b). Whilst the zooarchaeological and other archaeological data are still limited, they do suggest that fundamental changes in settlement and subsistence were taking place as early as the later 6th to early 7th century, before the establishment and growth of the *emporia*.

Archaeologists have tended to look at the Early Saxon period as a unified whole: non-Christian, non-urban, and essentially prehistoric. The vast majority of our information about Early Saxon England comes from cemeteries rather than settlement sites. Using dating techniques such as dendrochronology, stratigraphic analyses, and possibly AMS radiocarbon dating, we need to subdivide the Early Saxon period more finely. We also need more archaeozoological data from well-dated Early Saxon sites in eastern England. Sites like West Stow that were abandoned during the long 8th century may not provide the best data for studying the critical period of social, political and economic change. We also need to look at sites such as Quarrington and West Heslerton that continued to be occupied throughout the Early and Middle Saxon periods, and sites that were founded in the late 6th and 7th centuries but were occupied into the Middle Saxon era. By studying changes in animal husbandry practices within the Early Saxon period, we can begin to understand the economic changes that underlay the foundation of the *emporia* and the emergence of the Anglo-Saxon kingdoms.

	No.
Domestic Mammals	
Cattle (*Bos taurus*)	595
Sheep (*Ovis aries*)	15
Goat (*Capra hircus*)	13
Sheep/Goat	431
Pig (*Sus scrofa*)	952
Horse (*Equus caballus*)	5
Dog (*Canis familiaris*)	2
Cat (*Felis catus*)	6
Wild Mammals	
Red deer (*Cervus elaphus*)	1
Roe deer (*Capreolus capreolus*)	1
Domestic Birds	
Domestic fowl (*Gallus gallus*)	159
Domestic duck/Mallard (*Anas platyrhynchos*)	1
Wild Birds	
Jackdaw (*Corvus monedula*)	3
Total Identified Specimens	**2184**

Table A.9 Species List for Wingfield Street/Foundation Street Ipswich (4601)

	No.
Domestic Mammals	
Cattle (*Bos taurus*)	163
Sheep (*Ovis aries*)	1
Goat (*Capra hircus*)	1
Sheep/Goat	146
Pig (*Sus scrofa*)	210
Dog (*Canis familiaris*)	1
Cat (*Felis catus*)	4
Wild Mammals	
Red deer (*Cervus elaphus*)	2
Roe deer (*Capreolus capreolus*)	1
Domestic Birds	
Domestic fowl (*Gallus gallus*)	33
Domestic goose (*Anser anser*)	12
Total Identified Specimens	**574**

Table A.11 Species List for Buttermarket/St Stephen's Lane Ipswich (3104)

	No.
Domestic Mammals	
Cattle (*Bos taurus*)	1842
Sheep (*Ovis aries*)	14
Goat (*Capra hircus*)	18
Sheep/Goat	720
Pig (*Sus scrofa*)	1046
Horse (*Equus caballus*)	40
Dog (*Canis familiaris*)	4
Cat (*Felis catus*)	24
Wild Mammals	
Red deer (*Cervus elaphus*)	4
Roe deer (*Capreolus capreolus*)	10
Hare (*Lepus sp.*)	1
Otter (*Lutra lutra*)	1
Domestic Birds	
Domestic fowl (*Gallus gallus*)	149
Domestic goose (*Anser anser*)	56
Domestic duck/Mallard (*Anas platyrhynchos*)	7
Wild Birds	
Pigeons (Columbidae)	1
Raven (*Corvus corax*)	1
Crows (Corvidae)	1
Total Identified Specimens	**3939**

Table A.10 Species List for St Peter's Street Ipswich (5203)

Bibliography

Albarella, U. and Thomas, R., 2002 — 'They dined on crane: bird consumption, wild fowling and status in medieval England', *Acta Zoologica Cracovensia* 45, 23–38

Albarella, U., Johnstone, C. and Vickers, K., 2008 — 'The development of animal husbandry from the Late Iron Age to the end of the Roman period: a case study from south-east Britain', *J. Archaeol. Sci.* 35, 1828–48

Archer, S., 2003 — *The zooarchaeology of an Anglo-Saxon village: the faunal assemblage from Kilham, East Yorkshire*, unpublished MA thesis University of York

Baker, P., 2002 — *The Vertebrate Faunal Remains from Six Saxon Sites in the Lincolnshire and Norfolk Fenlands (Saxon Fenland Management Project)*, English Heritage Centre for Archaeology Report 46/2002

Bartosiewicz, L., Van Neer, W. and Lentacker, A., 1997 — *Draught Cattle: Their Osteological Identification and History* (Tervuren, Musée Royale de l'Afrique Centrale)

Baxter, I., 2010a — 'Small Roman dogs', in BoneCommons, Item #901, http://alexandriaarchive.org/bonecommons/items/show/901 (accessed July 20, 2010)

Baxter, I., 2010b — 'Animal bone', in: R. Atkins and A.Connor, *Farmers and Ironsmiths: Prehistoric, Roman and Anglo-Saxon Settlement beside Brandon Road, Thetford, Norfolk*, E. Anglian Archaeol. 134, 87–101

Binford, L.R., 1984 — *Faunal Remains from the Klasies River Mouth* (New York, Academic)

Blair, J., 2005 — *The Church in Anglo-Saxon Society* (Oxford University Press)

Blinkhorn, P., 1999 — 'Of cabbages and kings: production, trade and consumption in middle-Saxon England', in Anderton, M., (ed.), *Anglo-Saxon Trading Centres: Beyond the Emporia*, 4–23 (Glasgow, Cruithine)

Blackmore, L., 2002 — 'The origins of growth of Lundenwic, a mart of many nations', *Acta Archaeologica Lundensia* 8, 273–301

Boessneck, J. A., Müller, H.-H. and Tiechert, M., 1964 — 'Osteologische Unterschneidungsmerkmale zwischen Schaf (*Ovis aries* Linné) und Ziege (*Capra hircus* Linné)', *Kühn-Archiv* 78, 1–129

Bond, J., 1994 — 'Appendix I: the cremated animal bone', in J. McKinley, *The Anglo-Saxon Cemetery at Spong Hill, North Elmham, Part VIII: The Cremations*, E. Anglian Archaeol. 69, 122–35

Bond, J., 1995 — 'Animal bones from early Saxon sunken-featured buildings and pits', in R. Rickett, *The Anglo-Saxon Cemetery at Spong Hill, North Elmham, Part VII: the Iron Age, Roman and Early Saxon Settlement*, E. Anglian Archaeol. 73, 142–6

Bourdillon, J., 1988 — 'Countryside and town: the animal resources of Saxon Southampton', in Hooke, D. (ed.), *Anglo-Saxon Settlements*, 177–195 (Oxford, Blackwell)

Bourdillon, J. and Coy, J., 1977 — 'Statistical Appendix', to accompany 'The animal bones', in Holdsworth, P. (ed.), *Excavations at Melbourne Street, Southampton, 1971–76*, Counc. Brit. Archaeol. Res. Rep. 33, 79–121 (unpublished)

Bourdillon, J. and Coy, J., 1980 — 'The animal bones', in Holdsworth, P. (ed.), *Excavations at Melbourne Street, Southampton, 1971–76*, Counc. Brit. Archaeol. Res. Rep. 33, 79–121

Brain, C.K., 1967 — 'Hottentot food remains and their bearing on the interpretation of fossil bone assemblages', *Pap. Namib Desert Res. Station* 32

Brumfiel, E. and Earle, T.K., 1987 — 'Specialization, exchange, and complex societies: an introduction', in Brumfiel, E. and Earle, T. K. (eds.), *Specialization, Exchange, and Complex Societies*, 1–8 (Cambridge University Press)

Carr, R.D., Tester, A. and Murphy, P., 1988 — 'The Middle Saxon settlement at Staunch Meadow, Brandon', *Antiquity* 62, 371–7

Carver, M., 2005 — *Sutton Hoo: A Seventh-Century Burial Ground and its Context* (London, British Museum)

Chessel, D., Dufour, A.B. and Thioulouse, J., 2004 — 'The ade4 package – I: one table methods', *RNews* 4(1), 5–10

Childe, V.G., 1936 — *Man Makes Himself* (London)

Childe, V.G., 1950 — 'The urban revolution', *Town Planning Rev.* 21, 3–17

Clutton-Brock, J., 1976 — 'The animal resources', in Wilson, D.M. (ed.), *The Archaeology of Anglo-Saxon England*, 373–392 (London, Methuen)

Crabtree, P.J., 1982 — *Early Anglo-Saxon animal economy: an analysis of the animal bone remains from the early Saxon site of West Stow, Suffolk* (unpublished PhD thesis Univ. Pennsylvania)

Crabtree, P.J., 1984 — 'The archaeozoology of the Anglo-Saxon site at West Stow, Suffolk', in Biddick, K. (ed.), *Archaeological Approaches to Medieval Europe*, 223–235, Studies in Medieval Culture 18 (Kalamazoo, MI)

Crabtree, P.J., 1986 — 'Dairying in Irish prehistory: evidence from a ceremonial center, *Expedition* 28 (2), 59–62

Crabtree, P.J., 1989a — 'Sheep, horses, kine, and swine: a zooarchaeological approach to the Anglo-Saxon settlement of England', *J. Field Archaeol.* 16, 205–213

Crabtree, P.J., 1989b — 'Zooarchaeology at early Anglo-Saxon West Stow', in Redman, C. (ed.), *Medieval Archaeology*, 203–215, (Binghamton, NY, Medieval and Renaissance Texts and Studies)

Crabtree, P.J., 1990a — *West Stow, Early Anglo-Saxon Animal Husbandry*, E. Anglian Archaeol. 47

Crabtree, P.J., 1990b — 'Faunal remains', in West, S.E., *West Stow: the Prehistoric and Romano-British Occupations*, E. Anglian Archaeol. 48, 101–105

Crabtree, P.J., 1991 — 'Roman Britain to Anglo-Saxon England: the zooarchaeological evidence', in Crabtree, P.J. and Ryan, K. (eds.), *Animal Use and Culture Change*, MASCA Research Papers in Science and Archaeology, Supplement to Vol. 8, 32–38 (Philadelphia)

Crabtree, P.J., 1993 — 'The economy of an early Anglo-Saxon village: zooarchaeological research at West Stow', in P. Bogucki (ed.) *Case Studies in European*

Prehistory, 287–307 (Boca Raton, FL, CRC Press)

Crabtree, P.J., 1995 'The wool trade and the rise of urbanism in Middle Saxon England', in Wailes, B. (ed.), *Craft Specialization and Social Evolution: A Symposium in Commemoration of V. Gordon Childe*, 99–105 (Philadelphia, University Museum Publications)

Crabtree, P.J., 1996 'Production and consumption in an early complex society: animal use in Middle Saxon East Anglia', *World Archaeol.* 28 (1), 58–75

Crabtree, P.J., 2007 'Animals as material culture in the Middle Ages: the zooarchaeological evidence for wool production at Brandon', in Pluskowski, A. (ed.), *Breaking and Shaping Beastly Bodies: Animals as Material Culture in the Middle Ages*, 161–168 (Oxford, Oxbow)

Crabtree, P.J., 2010a 'Zooarchaeology and colonialism in Roman Britain: evidence from Icklingham', in Campana, D., Crabtree, P., deFrance, S., Lev-Tov, J. and Choyke, A. (eds.), *Anthropological Approaches to Zooarchaeology: Colonialism, Complexity and Animal Transformations*, 190–194 (Oxford, Oxbow)

Crabtree, P., 2010b 'Agricultural innovation and socio-economic change in early medieval Europe: Evidence from Britain and France', *World Archaeol.* 42, 122–136

Crabtree, P., 2012a 'The diet of Ipswich from the Middle Saxon through the Medieval periods', in Choyke, A., Jaritz, G. and Pluskowski, A. (eds.), *Animaltown: Beasts in Medieval Urban Space* (Oxford, Oxbow)

Crabtree, P.J., 2012b 'The animal bones from the Chapter House at St Albans Abbey', in Biddle, M. and Kjølbye-Biddle, B., *The Chapter House at St Albans Abbey*, Fraternity Friends St Albans Abbey Monogr. 1 (Oxford)

Crabtree, P. and Campana, D., 1987 'ANIMALS—a C language computer program for the analysis of faunal remains and its use in the analysis of the early Iron Age fauna from Dun Ailinne', *Archaeozoologia* 1, 58–69

Crabtree, P. and Campana, D., n.d. *The faunal remains from Brandon*, unpublished report for Suffolk County Council Archaeological Service, Bury St Edmunds

Crabtree, P. and Stevens, P., 1995 *Animal bones recovered from Wicken Bonhunt, Essex*, unpublished report for English Heritage

Davis, S., 2001 'The horse head from Grave 47', in W. Filmer-Sankey and T. Pestell, *Snape Anglo-Saxon Cemetery: Excavations and Surveys 1824–1992*, E. Anglian Archaeol. 95, 231–2

Dobney, K. and Jacques, D., 2002 'Avian signatures for identity and status in Anglo-Saxon England', *Acta Zoologica Cracoviensia* 45, 7–21

Dobney, K., Jacques, D., Barrett, J. and Johnstone, C., 2007 *Farmers, Monks and Aristocrats: the Environmental Archaeology of Anglo-Saxon Flixborough* (Oxford, Oxbow)

Done, G., 1993 'Animal bones from Anglo-Saxon contexts', in H. Hamerow, *Excavations at Mucking Volume 2: the Anglo-Saxon Settlement*, 74–79 (London, English Heritage)

Evans, C. and Serjeantson, D., 1988 'The backwater economy of a fen-edge community in the Iron Age: Upper Delphs, Haddenham', *Antiquity* 62 (235), 360–70

Evans, E.-J., 2007 'Animal bone', in A. Hardy, B.M. Charles and R.J. Williams, *Death and Taxes: the archaeology of a Middle Saxon estate centre at Higham Ferrers, Northamptonshire*, 145–157 (Oxford Archaeology)

Fern, C., 2007 'Early Anglo-Saxon horse burials from the fifth to the seventh centuries AD', *Anglo-Saxon Stud. Archaeol. Hist.* 14, 92–109

Filmer-Sankey, W. and Pestell, T., 2001 *Snape Anglo-Saxon Cemetery: Excavations and Surveys 1824–1992*, E. Anglian Archaeol. 95

Gardiner, M., 1997 'The exploitation of sea mammals in medieval England: bones and their social context', *Archaeol. J.* 154, 173–95

Grant, A., 1982 'The use of tooth wear as a guide to the ageing of the domestic ungulates', in Wilson, B., Grigson, C. and Payne, S. (eds), *Ageing and Sexing Animal Bones from Archaeological Sites*, Brit. Archaeol. Rep. British Ser. 109, 91–108 (Oxford)

Halstead, P., Collins, P. and Isaakidou, K., 2002 'Sorting the sheep from the goats: morphological distinctions between the mandibles and mandibular teeth of *Ovis* and *Capra*', *J. Archaeol. Sci.* 29, 545–553

Hamerow, H., 1993 *Excavations at Mucking, vol. 2: The Anglo-Saxon Settlement*, English Heritage Archaeol. Rep. 21

Hansen, I.L. and Wickham, C. (eds.), 2000 *The Long Eighth Century* (Leiden, Brill)

Harcourt, R., 1974 'The dog in prehistoric and early historic Britain', *J. Archaeol. Sci.* 1, 151–76

Harcourt, R., 1979 'The animal bones', in Wainwright, G. (ed.), *Gussage All Saints: An Iron Age Settlement in Dorset*, Dept. Environ. Archaeol. Rep. 10, 150–60

Hardy, A., Charles, B.M. and Williams, R.J., 2007 *Death and Taxes: the archaeology of a Middle Saxon estate centre at Higham Ferrers, Northamptonshire* (Oxford, Oxbow)

Haslam, J., 1981 'A Middle Saxon iron-working site at Ramsbury, Wilts.', *Medieval Archaeol.* 24, 1–68

Heinzel, H., Fitter, R. and Parslow, J., 1972 *The Birds of Britain and Europe* (London, Collins)

Higbee, L., 2005 'Large vertebrates', in R. Mortimer, R. Regan and S. Lucy, *The Saxon and Medieval Settlement at West Fen Road, Ely: The Ashwell site*, E. Anglian Archaeol. 110, 89–96

Higbee, L., 2009 'Mammal and bird bone', in S. Lucy, J. Tipper and A. Dickens, *The Anglo-Saxon Settlement at Bloodmoor Hill, Carlton Colville, Suffolk*, E. Anglian Archaeol. 131, 279–304

Higgs, E.S. (ed.), 1975 *Paleoeconomy* (Cambridge University Press)

Hinton, D.A., 1990 *Archaeology, Economy and Society: England from the Fifth to the Fifteenth Century* (London, Seaby)

Hodges, R., 1982 *Dark Age Economics* (London, Duckworth)

Hodges, D. and Hobley, B. (eds.), 1988 *The Rebirth of Towns in the West, AD 700–1050*, Counc. Brit. Archaeol. Res. Rep. 68

Hooke, D., 1998 *The Landscape of Anglo-Saxon England* (London, Leicester University Press)

Howard, M., 1963   'An assessment of a prehistoric technique of bovine husbandry', in Mournat, A.E. and Zeuner, F.E. (eds.), *Man and Cattle*, Roy. Anthrop. Inst. Occas. Pap. 18, 92–100

Hunter-Mann, K., 2001   *Lowthorpe Beck, Kilham, East Yorkshire, Report on the Training Excavation (Second Season), 2001* (York, York Archaeological Trust)

Ingram, C., 2001   'The animal bones from Oxford Science Park, Oxford', in J. Moore, 'Excavations at Oxford Science Park, Littlemore, Oxford', *Oxoniensia* 66, 163–219

Jones, R.T., n.d.   'Ancient Monuments Laboratory computer based osteometry data capture computer user manual', Ancient Monuments Laboratory Report No. 3342

Lascelles, G., 1892   'Falconry', in Cox, H. and Lascelles, G., *Coursing and Falconry*, 217–371 (Boston, Little, Brown)

Levine, M., Whitwell, K. and Jeffcott, L., 2002   'A Romano-British horse burial from Icklingham, Suffolk', *Archaeofauna* 11, 63–102

Loveluck, C., 2007   *Rural Settlement, Lifestyles and Social change in the Later First Millennium AD: Anglo-Saxon Flixborough in Its Wider Context* (Oxford, Oxbow)

Loveluck, C. and Atkinson, D., 2007   *The Early Medieval Settlement Remains from Flixborough, Lincolnshire: the Occupation Sequence, c. AD 600–1000* (Oxford, Oxbow)

Loveluck, C. and Tys, D., 2006   'Coastal societies, exchange and identity along the Channel and the southern North Sea shores of Europe, AD 600–1000', *J. Maritime Archaeology* 1, 140–69

Lucy, S., Tipper, J. and Dickens, A., 2009   *The Anglo-Saxon Settlement at Bloodmoor Hill, Carlton Colville, Suffolk*, E. Anglian Archaeol. 131

Luff, R., 1993   *Animal bones from excavations in Colchester, 1971–85*, Colchester Archaeol. Rep. 12 (Colchester Archaeological Trust)

Lyman, R.L., 1994   *Vertebrate Taphonomy* (Cambridge University Press)

Lyman, R.L., 2008   *Quantitative Paleozoology* (Cambridge University Press)

MacKinnon, M., 2010a   'Cattle "breed" variation and improvement in Roman Italy: connecting the zooarchaeological and ancient textual evidence', *World Archaeology* 42, 55–73

MacKinnon, M., 2010b   ' "Tails" of Romanization: animals and inequality in the Roman Mediterranean context', Paper presented at the 2010 Meeting of the Society for American Archaeology, April 2010, St Louis, Missouri, USA

Maltby, M., 1981   'Iron Age, Romano-British and Anglo-Saxon animal husbandry—a review', in Jones, M. and Dimbleby, G. (eds.), *The Environment of Man: the Iron Age to the Anglo-Saxon Period*, Brit. Archaeol. Rep. 87 (Oxford), 155–203

Maltby, M., 2002   'Animal bones in archaeology: how archaeozoologists can make a greater contribution to Iron Age and Romano-British archaeology', in K. Dobney and T. O'Connor (ed.), *Bones and the Man: Studies in Honour of Don Brothwell*, 88–94 (Oxford, Oxbow)

Maltby, M., 2006   'Salt and animal products: linking production and use in Iron Age Britain', in M. Maltby (ed.), *Integrating Zooarchaeology*, 117–122 (Oxford, Oxbow)

Maltby, M., 2010   *Feeding a Roman Town: Environmental evidence from excavations in Winchester, 1972–1985* (Winchester, Winchester Museums)

Maltby, M. n.d., a   *Oakley Road Clapham Animal Bones*, unpublished report available from the author

Maltby, M. n.d., b   *Analysis of Animal Bones from the Early to Middle Saxon Deposits (Period 4) from Harrold Bedfordshire*, unpublished report available from the author

Marzluff, J.M. and Angell, T., 2005   *In the Company of Crows and Ravens* (New Haven, Yale University Press)

McCormick, F., 1992   'Early faunal evidence for dairying', *Oxford J. Archaeol.* 11, 201–209

McCormick, F., 2008   'The decline of the cow: agricultural and settlement change in early medieval Ireland', *Peritia* 20, 210–225

McCormick, F. and Murray, E., 2007   *Knowth and the Zooarchaeology of Early Christian Ireland* (Dublin, Royal Irish Academy)

Milne, C. and Crabtree, P., 2001   'Prostitutes, a rabbi, and a carpenter — dinner at the Five Points in the 1980s', *Hist. Archaeol.* 33 (2), 31–48

Morant, P., 1758   *History and Antiquities of the County of Essex* (London, T. Osborne)

Moreland, J., 2000   'The significance of production in eighth-century England', in Hansen, I.L. and Wickham, C. (ed.), 69–104 (Leiden, Brill)

Mulville, J. 2003   'Phases 2a–2e: Anglo-Saxon occupation', in A. Hardy, A. Dodd and G.C. Keevill, *Aelfric's Abbey: Excavations at Eynsham Abbey, Oxfordshire 1989–92*, 343–360 (Oxford Archaeological Unit)

Mulville, J. and Ayres, K., 2004   'Chapter 18: animal bones', in G. Hey, *Yarnton: Saxon and Medieval Settlement and Landscape*, 325–350 (Oxford Archaeological Unit)

Mulville, J., Bond, J. and Craig, O., 2005   'The white stuff, milking in the Outer Scottish Isles', in J. Mulville and A.K. Outram (ed.), *The Zooarchaeology of Fats, Oils, Milk and Dairying*, 167–182 (Oxford, Oxbow)

O'Connor, T.P., 1991   *Bones from 46–54 Fishergate*, The Archaeology of York 15/4 (London, Council for British Archaeology)

O'Connor, T.P., 1992   'Pets and pests in Roman and medieval Britain', *Mammal Review* 22, 107–113

O'Connor, T.P., 2010   'Livestock and deadstock in early medieval Europe from the North Sea of the Baltic', *Environ. Archaeol.* 15, 1–15

Oosthuizen, S., 2007   'The Anglo-Saxon kingdom of Mercia and the origin and distribution of common fields', *Agricultural History Review* 55, 153–80

Payne, S., 1973   'Kill-off Patterns in Sheep and Goats: the Mandibles from Aşvan Kale', *Anatolian Studies* 23, 281–303

Powell, A. and Clark, K.M., 2002   'Animal bones', in A. Mudd, *Excavations at Melford Meadows, Brettenham, 1994: Romano-British and Early Saxon Occupations*, E. Anglian Archaeol. 99, 101–8

Prummel, W., 1983   *Excavations at Dorestad 2: Early Medieval Dorestad an Archaeozoological Study* (Amersfoort, ROB)

Rackham, J., 2003   'Animal bones', in G. Taylor, 'An early to middle Saxon settlement at Quarrington', *Antiq. J.* 83, 258–73

Reynolds, A., 2005   'On farmers, traders and kings: archaeological reflections of social complexity in early medieval north western Europe', *Early Medieval Europe* 13, 97–118

Rielly, K., 2003   'The animal and fish bone', in Malcolm, G. and Bowsher, D., *Middle Saxon London: Excavations at the Royal Opera House 1989–1999*, Mus.London Archaeol. Serv. Monogr. 15, 315–324

Roberts, T., 2005   'Animal bone', in J. Murray and T. MacDonald, 'Excavations at Station Road, Gamlingay, Cambridgeshire', *Anglo-Saxon Stud. Archaeol. Hist.* 13, 246–8

Rothera, S., ed., 1998   *Breckland Natural Area Profile, Natural England (Norwich)*, available online: http://www.english-nature.org.uk/science/natural/profiles%5cnaProfile46.pdf

Royal Society for the Protection of Birds, 2007   *Carrot fields to cranes in 11 years*, available online: http://www.birdguides.com/webzine/article.asp?a=1009 (accessed: 17 January 2009)

Scull, C., 2009   *Early Medieval (Late 5th–Early 8th Centuries AD) Cemeteries at Boss Hall and Buttermarket, Ipswich, Suffolk*, Soc. Medieval Archaeol. Monogr. 27 (Leeds)

Serjeantson, D., 2009   *Birds* (Cambridge University Press)

Sherratt, A.G., 1981   'Plough and pastoralism: aspects of the secondary products revolution', in Hodder, I., Isaac, G. and Hammond, N. (ed.), *Pattern of the Past: Studies in Honour of David Clarke*, 261–305 (Cambridge University Press)

Sherratt, A.G., 1983   'The secondary exploitation of animals in the Old World', *World Archaeol.* 15, 90–104

Silver, I.I., 1969   'The ageing of the domestic mammals', in Brothwell, D. and Higgs, E.S. (eds.), *Science in Archaeology*, 283–302 (London, Thames and Hudson)

Solti, B., 1985   'Vergleichende osteometrische Untersuchungen uber den Korperbau europaischer Grossfalken sowie dessen funktionelle Beziehungen', *Folia Historico Naturalia Musei Matraensis* 10, 115–30

Stevens, P., 1994   *Wicken Bonhunt, Essex: the animal bone from the Romano-British to the Saxo-Norman periods*, unpublished report for English Heritage

Sykes, N., 2004   'The dynamics of status symbols: wildfowl exploitation in England AD 410–1550', *Archaeol. J.* 161, 82–105

Taylor, G., 2003   'An early to middle Saxon settlement at Quarrington, Lincolnshire', *Antiq. J.* 83, 231–80

Tiechert, K., 1984   'Size variation in cattle from Germania Romana and Germania Libera', in Grigson, C. and Clutton-Brock, J. (eds.), *Animals in Archaeology 4: Animal Husbandry in Europe*, Brit. Archaeol. Rep. Int. Ser. 227, 93–103 (Oxford)

Tyler, N.J.C., 1987   'Sexual dimorphism in the pelvic bones of Svalbard reindeer, *Rangifer tarandus platyrhychus*', *J. Zoology* 213, 147–52

Vandervell, A. and Coles, C., 1980   *Game and the English Landscape: the Influence of the Chase on Sporting Art and Scenery* (New York, Viking)

von den Driesch, A., 1976   *A Guide to the Measurement of Animal Bones from Archaeological Sites*, Peabody Mus. Bull. 1 (Harvard University)

von den Driesch, A. and Boessneck, J., 1974   'Kritische Anmerkungen zur Widerristhöhenberechnung aus Längenmassen vor- und frühgeschichtlicher Tierknocken', *Säugetierkunliche Mitteilungen* 22, 325–48

Wade, K., 1980   'A settlement site at Bonhunt Farm, Wicken Bonhunt, Essex', in Buckley, D.G. (ed.), *Archaeology in Essex to AD1500*, Counc. Brit. Archaeol. Res. Rep. 34, 96–102 (London)

Wade, K., 1988   'Ipswich', in Hodges, R. and Hobley, B. (ed.), *The Rebirth of Towns in the West, AD 700–1050*, Counc. Brit. Archaeol. Res. Rep. 68, 93–100

Wade, K., 1993   'The urbanisation of East Anglia: the Ipswich perspective', in Gardiner, J. (ed.), *Flatlands and Wetlands: Current Themes in East Anglian Archaeology*, E. Anglian Archaeol. 50, 144–151

Wade, K., 2000   'Ipswich', in Crabtree, P. (ed.), *Medieval Archaeology: an Encyclopedia*, 173–175 (New York, Garland)

Watkins, A., 2010   'Aelfric's Colloquy, translated from the Latin', *Kent Archaeology* 16, 1–14

West, B., 1989   'Birds and mammals from the Peabody Site and National Gallery', in Whitehead, R. L. and Cowie, R., 'Excavations at the Peabody Site, Chandos Place and the National Gallery', *Trans. London Middlesex Archaeol. Soc.* 40, 150–160

West, S.E., 1985   *West Stow, the Anglo-Saxon Village*, E. Anglian Archaeol. 24

West, S.E., 1990   *West Stow, the Prehistoric and Roman Occupations*, E. Anglian Archaeol. 48

West, S.E. and Plouviez, J., 1976   'The Roman site at Icklingham', E. Anglian Archaeol. 3, 63–125

White, R. and Barker, P., 1998   *Wroxeter, Life and Death of a Roman City*, (Stroud, Tempus)

Wilson, T., 1995   'Animal bones', in P. Andrews, *Excavations at Redcastle Furze, Thetford, 1988–9*, E. Anglian Archaeol. 72, 121–8

Yvenic, J.-H. and Chaulet, V., 2009   'Évolution des choix d'levage durant le Haut Moyen Age, dans le Nord de la France', Poster presented at Archeometrie 2009, Montpelier, 6–10 April

Zeder, M.A., 1988   'Understanding urban process through the study of specialized subsistence economy in the Near East', *J. Anthropol. Archaeol.* 7, 1–55

Zeder, M.A., 1991   *Feeding Cities: Specialized Animal Economy in the Ancient Near East* (Washington, Smithsonian)

Index

Illustrations are denoted by page numbers in *italics*.

East Anglian Archaeology

is a serial publication sponsored by ALGAO EE and English Heritage. It is the main vehicle for publishing final reports on archaeological excavations and surveys in the region. For information about titles in the series, visit **www.eaareports.org.uk**. Reports can be obtained from:

Oxbow Books, 10 Hythe Bridge Street, Oxford OX1 2EW

or directly from the organisation publishing a particular volume.

Reports available so far:

No.1, 1975 Suffolk: various papers
No.2, 1976 Norfolk: various papers
No.3, 1977 Suffolk: various papers
No.4, 1976 Norfolk: Late Saxon town of Thetford
No.5, 1977 Norfolk: various papers on Roman sites
No.6, 1977 Norfolk: Spong Hill Anglo-Saxon cemetery, Part I
No.7, 1978 Norfolk: Bergh Apton Anglo-Saxon cemetery
No.8, 1978 Norfolk: various papers
No.9, 1980 Norfolk: North Elmham Park
No.10, 1980 Norfolk: village sites in Launditch Hundred
No.11, 1981 Norfolk: Spong Hill, Part II: Catalogue of Cremations
No.12, 1981 The barrows of East Anglia
No.13, 1981 Norwich: Eighteen centuries of pottery from Norwich
No.14, 1982 Norfolk: various papers
No.15, 1982 Norwich: Excavations in Norwich 1971–1978; Part I
No.16, 1982 Norfolk: Beaker domestic sites in the Fen-edge and East Anglia
No.17, 1983 Norfolk: Waterfront excavations and Thetford-type Ware production, Norwich
No.18, 1983 Norfolk: The archaeology of Witton
No.19, 1983 Norfolk: Two post-medieval earthenware pottery groups from Fulmodeston
No.20, 1983 Norfolk: Burgh Castle: excavation by Charles Green, 1958–61
No.21, 1984 Norfolk: Spong Hill, Part III: Catalogue of Inhumations
No.22, 1984 Norfolk: Excavations in Thetford, 1948–59 and 1973–80
No.23, 1985 Norfolk: Excavations at Brancaster 1974 and 1977
No.24, 1985 Suffolk: West Stow, the Anglo-Saxon village
No.25, 1985 Essex: Excavations by Mr H.P.Cooper on the Roman site at Hill Farm, Gestingthorpe, Essex
No.26, 1985 Norwich: Excavations in Norwich 1971–78; Part II
No.27, 1985 Cambridgeshire: The Fenland Project No.1: Archaeology and Environment in the Lower Welland Valley
No.28, 1985 Norfolk: Excavations within the north-east bailey of Norwich Castle, 1978
No.29, 1986 Norfolk: Barrow excavations in Norfolk, 1950–82
No.30, 1986 Norfolk: Excavations at Thornham, Warham, Wighton and Caistor St Edmund, Norfolk
No.31, 1986 Norfolk: Settlement, religion and industry on the Fen-edge; three Romano-British sites in Norfolk
No.32, 1987 Norfolk: Three Norman Churches in Norfolk
No.33, 1987 Essex: Excavation of a Cropmark Enclosure Complex at Woodham Walter, Essex, 1976 and An Assessment of Excavated Enclosures in Essex
No.34, 1987 Norfolk: Spong Hill, Part IV: Catalogue of Cremations
No.35, 1987 Cambridgeshire: The Fenland Project No.2: Fenland Landscapes and Settlement, Peterborough–March
No.36, 1987 Norfolk: The Anglo-Saxon Cemetery at Morningthorpe
No.37, 1987 Norfolk: Excavations at St Martin-at-Palace Plain, Norwich, 1981
No.38, 1987 Suffolk: The Anglo-Saxon Cemetery at Westgarth Gardens, Bury St Edmunds
No.39, 1988 Norfolk: Spong Hill, Part VI: Occupation during the 7th–2nd millennia BC
No.40, 1988 Suffolk: Burgh: The Iron Age and Roman Enclosure
No.41, 1988 Essex: Excavations at Great Dunmow, Essex: a Romano-British small town in the Trinovantian Civitas
No.42, 1988 Essex: Archaeology and Environment in South Essex, Rescue Archaeology along the Gray's By-pass 1979–80
No.43, 1988 Essex: Excavation at the North Ring, Mucking, Essex: A Late Bronze Age Enclosure
No.44, 1988 Norfolk: Six Deserted Villages in Norfolk
No.45, 1988 Norfolk: The Fenland Project No. 3: Marshland and the Nar Valley, Norfolk
No.46, 1989 Norfolk: The Deserted Medieval Village of Thuxton
No.47, 1989 Suffolk: West Stow: Early Anglo-Saxon Animal Husbandry
No.48, 1989 Suffolk: West Stow, Suffolk: The Prehistoric and Romano-British Occupations

No.49, 1990 Norfolk: The Evolution of Settlement in Three Parishes in South-East Norfolk
No.50, 1993 Proceedings of the Flatlands and Wetlands Conference
No.51, 1991 Norfolk: The Ruined and Disused Churches of Norfolk
No.52, 1991 Norfolk: The Fenland Project No. 4, The Wissey Embayment and Fen Causeway
No.53, 1992 Norfolk: Excavations in Thetford, 1980–82, Fison Way
No.54, 1992 Norfolk: The Iron Age Forts of Norfolk
No.55, 1992 Lincolnshire: The Fenland Project No.5: Lincolnshire Survey, The South-West Fens
No.56, 1992 Cambridgeshire: The Fenland Project No.6: The South-Western Cambridgeshire Fens
No.57, 1993 Norfolk and Lincolnshire: Excavations at Redgate Hill Hunstanton; and Tattershall Thorpe
No.58, 1993 Norwich: Households: The Medieval and Post-Medieval Finds from Norwich Survey Excavations 1971–1978
No.59, 1993 Fenland: The South-West Fen Dyke Survey Project 1982–86
No.60, 1993 Norfolk: Caister-on-Sea: Excavations by Charles Green, 1951–55
No.61, 1993 Fenland: The Fenland Project No.7: Excavations in Peterborough and the Lower Welland Valley 1960–1969
No.62, 1993 Norfolk: Excavations in Thetford by B.K. Davison, between 1964 and 1970
No.63, 1993 Norfolk: Illington: A Study of a Breckland Parish and its Anglo-Saxon Cemetery
No.64, 1994 Norfolk: The Late Saxon and Medieval Pottery Industry of Grimston: Excavations 1962–92
No.65, 1993 Suffolk: Settlements on Hill-tops: Seven Prehistoric Sites in Suffolk
No.66, 1993 Lincolnshire: The Fenland Project No.8: Lincolnshire Survey, the Northern Fen-Edge
No.67, 1994 Norfolk: Spong Hill, Part V: Catalogue of Cremations
No.68, 1994 Norfolk: Excavations at Fishergate, Norwich 1985
No.69, 1994 Norfolk: Spong Hill, Part VIII: The Cremations
No.70, 1994 Fenland: The Fenland Project No.9: Flandrian Environmental Change in Fenland
No.71, 1995 Essex: The Archaeology of the Essex Coast Vol.I: The Hullbridge Survey Project
No.72, 1995 Norfolk: Excavations at Redcastle Furze, Thetford, 1988–9
No.73, 1995 Norfolk: Spong Hill, Part VII: Iron Age, Roman and Early Saxon Settlement
No.74, 1995 Norfolk: A Late Neolithic, Saxon and Medieval Site at Middle Harling
No.75, 1995 Essex: North Shoebury: Settlement and Economy in South-east Essex 1500–AD1500
No.76, 1996 Nene Valley: Orton Hall Farm: A Roman and Early Anglo-Saxon Farmstead
No.77, 1996 Norfolk: Barrow Excavations in Norfolk, 1984–88
No.78, 1996 Norfolk:The Fenland Project No.11: The Wissey Embayment: Evidence for pre-Iron Age Occupation
No.79, 1996 Cambridgeshire: The Fenland Project No.10: Cambridgeshire Survey, the Isle of Ely and Wisbech
No.80, 1997 Norfolk: Barton Bendish and Caldecote: fieldwork in south-west Norfolk
No.81, 1997 Norfolk: Castle Rising Castle
No.82, 1998 Essex: Archaeology and the Landscape in the Lower Blackwater Valley
No.83, 1998 Essex: Excavations south of Chignall Roman Villa 1977–81
No.84, 1998 Suffolk: A Corpus of Anglo-Saxon Material
No.85, 1998 Suffolk: Towards a Landscape History of Walsham le Willows
No.86, 1998 Essex: Excavations at the Orsett 'Cock' Enclosure
No.87, 1999 Norfolk: Excavations in Thetford, North of the River, 1989–90
No.88, 1999 Essex: Excavations at Ivy Chimneys, Witham 1978–83
No.89, 1999 Lincolnshire: Salterns: Excavations at Helpringham, Holbeach St Johns and Bicker Haven
No.90, 1999 Essex:The Archaeology of Ardleigh, Excavations 1955–80
No.91, 2000 Norfolk: Excavations on the Norwich Southern Bypass, 1989–91 Part I Bixley, Caistor St Edmund, Trowse
No.92, 2000 Norfolk: Excavations on the Norwich Southern Bypass, 1989–91 Part II Harford Farm Anglo-Saxon Cemetery
No.93, 2001 Norfolk: Excavations on the Snettisham Bypass, 1989
No.94, 2001 Lincolnshire: Excavations at Billingborough, 1975–8
No.95, 2001 Suffolk: Snape Anglo-Saxon Cemetery: Excavations and Surveys
No.96, 2001 Norfolk: Two Medieval Churches in Norfolk
No.97, 2001 Cambridgeshire: Monument 97, Orton Longueville